For my children Matt, Rae and Rob

HEADS UP
GENTLEMEN

A life of action and adventure
in the Australian Army

BEN LANS

First published in 2019 by Barrallier Books Pty Ltd,

trading as Echo Books

Registered Office: 35—37 Gordon Avenue, West Geelong, Victoria 3220, Australia.

www.echobooks.com.au

Copyright ©Ben Lans

Creator: Lans, Ben.

Title: /Heads up Gentlemen—A life of action and adventure
in the Australian Army/Ben Lans.

ISBN: 9780648355298 (softcover)

NATIONAL
LIBRARY
OF AUSTRALIA

A catalogue record for this
book is available from the
National Library of Australia

Book layout and design by Peter Gamble, Canberra.

Cover graphics and design by Ian Shiel.

Set in Garamond Premier Pro Display, 12/17 and/Trajan Pro.

www.echobooks.com.au

TABLE OF CONTENTS

PREFACE

This is a memoir about people, places, action and adventure. From the Franklin River deep in south-west Tasmania to tropical Townsville in Far North Queensland, from the jungles of Malaysia to the bitter face of the Berlin Wall, this memoir reflects the many facets of a satisfying career in the Australian Army.

This is about my Army experiences from the late 60's to the early 90's. I joined the Army as an officer cadet with little military background and a vague ambition to work with men and experience adventure. This is an account of personal experiences, emotions and humour in times of action and adventure at home, overseas and on operations; of loyalty and mateship developed through shared achievements and hardships.

These memoirs share my progress from a young and inexperienced officer sent to serve on operations in South Vietnam, into a leader and training developer. They describe the special and the mundane of service life, the fun, the reward and sometimes sadness but mostly satisfaction of working with soldiers. Postings include the Commonwealth Forces in Singapore in the days immediately after its independence. I describe the experiences gained during those post-colonial days of Singapore, in the early 1970s, the dawn of the new Asian age, and during training in the deep jungles of the Malay Peninsula, where elephants and tigers survived to threaten the unwary soldier. Later, as an artillery analyst in the mid-1980s, seconded and fully accredited to work for the UK Ministry of Defence in London as principal adviser on Soviet and Warsaw Pact artillery to the UK Ministry of Defence, I describe my role as part of the intriguing world that was the western intelligence community during the Cold War.

As a strong believer in the concept of adventurous training, I devised, promoted and executed many adventurous training exercises, some of which helped shape the concept of this kind of training in the Australian Army. These memoirs reflect the scale, danger, risks and satisfaction associated with the planning and execution of major expeditions, including a detailed and riveting account of adventure and survival on Tasmania's Franklin River, in 1980, before the river became well known to adventurers and world famous for its environmental significance.

There are first-hand accounts of working with, and learning from, other officers and soldiers whom I had the pleasure to know during my regimental, instructional and staff postings at home and abroad.

1 Heads up Gentlemen

I stand on parade, rigid, hands clasped behind my back, legs slightly apart. 'At ease', I think they call this. This is my first day in the Australian Army. I'm at the Officer Cadet School, in Portsea, on the Mornington Peninsula of Port Phillip Bay in Victoria. It's a mid-winter's day and it's not yet dawn. I'm a Queenslander but I was born in Holland and I've long forgotten just how cold it can get, anywhere, in Holland or here at the bottom end of Australia. The icy morning air on my face comes as a shock.

From out of the darkness a calm and reassuring voice is telling me I must stand straight, stand still, and keep my head up, and the voice addresses us all as gentlemen.

'Heads up gentlemen!'

I push out my chest. Never have I been called a gentleman before! I become aware of the outlines of buildings around me, buildings that didn't register when I arrived yesterday. They surround the small parade ground and seem to be a mixture of barracks and old structures with verandas. There's a gap between the buildings to my front framing the glow of the rising sun. It's reflecting over the water. The beauty of it registers, though only for a moment!

There's the voice again. I want to find the owner, but I can't seem to take my eyes off that distant glow. Besides, I'm not game to move my head, even a fraction. And anyway, I think my eyeballs are frozen! I daren't move 'cause any movement against these starched Army greens is bloody painful! Who decided we shouldn't wear jumpers or singlets with our starched greens and get up before the sun?

But I take solace in that voice. I can see him now. A tall man, with a firm, kind voice. The voice seems so mature and has such authority. Is he a cadet? Is he on the staff? Where are the staff? Apart from the arrival at the railway station, I've seen and met only senior cadets. Surely someone other than cadets run this place?

'Heads up gentlemen!'

The voice again, and I respond again. I'm strangely reassured by this authoritative yet warm voice. I'm trying to make out the faces around me. They barely registered on my conscience last night as I was too busy trying to listen to the orders and instructions being barked at me. Does everyone in the Army always shout? Well, no... obviously, they don't! That voice from the front is being delivered without malice, there is no shouting, yet the message it conveys is very effective. I must remember that.

Later, I learned that the voice belonged to a senior class cadet named John Guy and that he was the BSM, the Battalion Sergeant Major, the most senior of the senior class. Just his title sounded impressive!

This was June 1968 and we'd arrived in Melbourne from all over Australia. Bright young faces, full of confidence and cockiness. After all, hadn't we been told during the enlistment process that we are among the country's top 10% of eligible males? We were full of confidence. Bring it on!

The trip from Brisbane was eventful only in that it had given me the last opportunity to see my girlfriend Lo, before being inducted into this new world of the military. Here we were, eight 'chaps'—another new term I came to learn in the Army—from good ol' Brissy, all dressed up in suits and keen to impress anyone who cared to be impressed, especially any girls that happened to look sideways. The train had arrived in Sydney on a Saturday morning and we had a complete day to 'entertain ourselves'—another typical Army phrase with which I was to become familiar—before catching the next train to Melbourne that night. My plan was to meet up with Lo who happened to be playing hockey somewhere in North Sydney. My mistake was telling the other chaps!

'Excellent mate! If your girlfriend is playing hockey there must be team full of girls! Lans, we're coming with you, let's go!'

Damn it! That was the last thing I wanted, but there was no stopping them and so it was that the West Ryde 1st Grade Women's Hockey Team found themselves supported by a loud and rather well-dressed bunch of would-be Army Officer Cadets! Fortunately, the effect of the many beers consumed the night before had worn off sufficiently to moderate the enthusiasm just enough. However, we did raise a few eyebrows. We must have looked quite ridiculous, cheering from the sideline in suits at ten in the morning!

After the game, despite valiant and mostly over-the-top attempts, none of my new mates seemed to be successful in doing the big 'pick up', so I ended up in a cab with a bemused Lo and the ever-present entourage of suited blokes crowded around us. Lo must have wondered what she was getting into. She told me later that it was funny but a bit embarrassing. Anyway, my style was completely cramped by these hangers-on and any plans that I had for a nice day with my girlfriend were smashed! We dropped her off at the train station where she caught the train home and that was it! Back to the beer drinking with this gang of suits, sitting on the pier at the Manly ferry terminal singing about 'chundering' into the Pacific Ocean, or something like that.

Luckily, the next night's train journey was quieter with less beer, for me anyway. My limit had well and truly been reached and I wanted to sleep. What would this place called Portsea Officer Cadet School be like? Would I cope? Was I fit enough? For months I'd been running most days of the week to get my fitness up, doing squats and chin ups, but I was worried. What about the discipline and the famed 'bastardisation' that I'd heard of. A warning of what was to come surfaced briefly on arrival in Melbourne.

As we alighted the train at Spencer Street Station, simultaneously meeting up with the other young men getting off buses and trains that had carried them from their home states, an impeccably dressed Army man came over to us. Wow! This bloke looked impressive, calm and full of authority in his immaculate uniform. Some of the boys (it seems wrong to call them men) identified him as a 'Warrant Officer'. The Warrant Officer simply picked up his 'clue board', another term new to me, and read the first name out loud, starting at the top of the alphabet. Clearly it was a trap. Some poor fellow whose only sin was that his name started with 'A' simply said:

'Yeah, I'm here!'

The clue board was lowered very slowly, the eyes narrowed into a deliberate icy and prolonged stare, and in a cold monotone that silenced us all, the Army man, without emotion, delivered an edict directly into the brain of the offending chap:

'You... will address me as... 'Sir! Is that clear?'.

'Yes Sir!' rallied a chorus of voices in obedient agreement.

So, this is how it is to be then? When you are at the bottom of the rung, just say 'Sir'. We all got it.

This was the first commandment.

I consider myself to be very lucky to have been there at all. The military had always been my choice of career, but my first attempt two years previously had ended in disappointment. My application for the Royal Military College, Duntroon, failed in mid-stream, apparently because of a hearing impediment in my right ear. I was shattered. But my friend in the CMF told me about the Officer Cadet School at Portsea and encouraged me to try again. Portsea offered a one-year all-military course instead of the four-year arts, science or engineering degree course at Duntroon. But would my below-standard hearing keep me out again? It should have, but with the cockiness of youth and an unfounded confidence, I decided to apply anyway.

There followed a moment that shaped my military career and thus my life. At the Eastern Command Personnel Depot in Enoggera, a suburb of Brisbane, I once again completed the Army medical. This included the same hearing test as the one for Duntroon two years prior. I was sitting in the waiting room, along with a bunch of other hopefuls, when suddenly a rotund, ruddy-faced Sergeant walked in and asked for me by name.

'There seems to be something wrong with your hearing test son, follow me!' said the rotund Sergeant.

My heart sank. Here we go again, I thought, this will confirm that bloody hearing loss a second time. But to my surprise, he beckoned me over and did the test again. This time though, when he fitted me with the earphones and sat me down, I found myself facing him. In those days when there were no booths, the person being tested would be seated facing away from the operator to avoid cheating. But from this position I could see his hands and as I watched him press the buttons on the console in front of him, I simply indicated with my thumb that yep, I heard that one! I watched his fingers and indicated with my thumb every time, beep or no beep in my ear. All done in no time. Perfect results!

'That's better son,' he said, as he ushered me back to the waiting room.

No hearing loss and a piece of paper to prove it! To this day, I do not understand why that friendly, ruddy faced Sergeant had taken pity on me, or was it that he had taken a liking to me? Anyway, he managed to fake my medical for me. I can remember his face still. I wanted to hug him but that was probably not a good idea.

Medical completed, I was hopeful that I would do well at the subsequent selection board. There were twelve of us on that board. Twelve young hopefuls. A panel of senior officers were conducting boards every day for two weeks, just in Brisbane. I was told they would be conducting boards in major cities all over Australia and that there had already been a weeding out process just to get to this stage. Hmmm, the odds were not looking good. My strongest memories of the selection board are that I was conscious of not appearing too 'pushy' in the many group aptitude and initiative tests; that the entire board of senior officers were watchful as we all sat down to a formal lunch in the Victoria Barracks Officers Mess (obviously minding our manners was important) and that the full table of interviewing officers, some ten of them led by an Army phycologist, smiled at me as I sat on my lone chair to the front of their table. So I remember smiling back at them.

What else to do? The image of their smiling countenances as I left the room is still burned into my brain today. Somewhere in the back of my mind a flame of hope was lit that I must have done OK. I was the only one that passed that day's board and in total there were only about eight Queenslanders selected that year out of the hundreds that applied.

But there was one more hurdle to overcome. After being informed of my selection I needed to show the recruiting officer my educational qualifications. The minimum standard for entry was a high school leaving certificate, or a 'senior pass' as it was called in Queensland. My certificate was woeful, having failed some of the subjects. My heart was not in it after the failed Duntroon attempt. I claimed a pass in my Portsea application but quite deliberately didn't provide any evidence and continued to conveniently 'forget' the documentation every time I spoke to the recruiting officer. It was touch and go, but I got away with it.

At Spencer Street, the incident with the Warrant Officer was sobering, but it only dampened our high spirits a little. We, the unsuspecting remained upbeat. On the bus I found myself next to a very tall black man who turned out to be from Kenya. Later I realised that there were also a few New Zealand, Malay, Pilipino and Papua New Guinea cadets, along with the 70 or so Aussies. As we drove along the shores of Port Phillip Bay, I realised that the Kenyan, whose name was Silas, had never even seen a ship before. It hadn't occurred to me that there would be candidates from countries other than Australia. We all chatted freely to each other, but as the bus began to draw closer to Portsea, the conversations began to dry up. One by one we became aware that we were coming closer and closer to a place that we knew little about. One by one we fell silent.

As the buses filed through the gates we craned our necks. And then, there they were! Impeccably and identically uniformed young men: Officer Cadets, lines of them! Our bus stopped at the end of one of the lines. A cadet jumped on and motioned for us to get out. He was a tall man, very straight and very smart in his perfectly ironed uniform. Silently we sauntered off. Making any sort of noise seemed inappropriate let alone talking or answering anything at all, especially after that first serve we all witnessed at the station. I can clearly remember the way the cadets refused to look us in the eye. Standing to attention, they simply used hand signals and head nods to indicate which way to walk, where to put our bags, where to stand and wait. No eye contact! No words! No shouting! It was unnerving!

Then, when we were all off the buses, standing in a sort of collective, self-conscious gaggle, a volley of instructions suddenly bellowed forth. From no information at all to this was information overload! Each of us was being allocated to a senior classman. Who, where... me? I found myself shaking the hand of my 'father'. Father? I was to be his 'son'? What was that supposed to mean? My 'father' was the first senior classman to look me in the eye and I quickly understood that he was to be my guardian. Cadet Corporal Ian Farrant. His look was reserved though. There was not much committal in those eyes. I clearly remember his officers peak cap drawn low over his eyebrows. Smart cap, I had the time to reflect. Later I was to learn that it was a cap manufactured by a company called 'Herbert Johnson' and was imported from England, and that no senior classman would be seen dead in an Australian Army issued officers peak cap. Buying a 'Herbie' from England was the go. To this day I look for and identify Herbert Johnson caps versus government issued caps in movies about Commonwealth armies.

My feeling of nervous anticipation dissipated a little and I relaxed. I felt I could trust him. Our fathers showed us our rooms and gave us run-downs on what was expected in every detail. Where to unpack our clothes; how to lay out every item from shirts to singlets and to ensure that the socks are neatly rolled and 'smile' at you when someone opens the drawer. A smiling set of socks is one where the final fold has the outer ends turned upwards, not downwards. My mum never told me about that one. We were divided into sections and platoons and companies and then my father led me down to another senior classman's room at the end of the corridor. He turned out to be the Platoon Sergeant. We filed in. I noticed immediately the homeliness of his room. Pictures of his family were on the desk. He was a relaxed and confident young man who chatted with us as if we were friends for life. He told us about what we should expect, and how to deal with the daily grind of the school. It didn't sound so bad really. Maybe these blokes were not so tough and maybe the bastardisation that I had been hearing about was not going to be all that bad. Maybe? Or was he playing the good cop/bad cop role... to gain our trust before the onslaught?

Right, off to the cadet's mess, a huge room in a beautiful old colonial building. A quick meal, some more vital instructions about timings and what to do when and where and what to expect, instructions that were fast overloading my brain, and then it was time for 'lights out'... another new expression, which I quickly learned meant bed! Relief at last! Relief from the onslaught of 'do this, do that, stand here, march there!'

In my small room, I felt quite alone, with only the light of the small bar heater on the wall and the silence of Port Phillip Bay just outside my window. But I soon learnt that my room was a place where I could get into my own head, where no-one else could bother me. It was pretty basic, with one wall containing built-in wardrobes and a built-in desk with an iron-frame bed pushed against the other wall. The window was one of those old-fashioned sash windows which had recently been painted, but the painters hadn't closed the window properly before applying the paint. The result was a permanent gap of about half an inch at the bottom that was letting in very cold air.

I stared at the ceiling and listened to the winter wind whistling through that crack under the window. This is cold I thought, really cold! What am I doing here? But there was no time to dwell on that now. Surprisingly, I soon fell into a deep sleep.

2 You Have a Big Head Son

A*gghh...what is that screeching?*

My brain scrambles as I become aware of my surroundings. It's pitch black outside, but they want us to get up before the sun does! That noise must be a recording! It's being trumpeted over a loudspeaker at maximum decibels, I reckon. This is rude! I don't get up in the dark! Reveille! A shock awakening for a young bloke. Where am I? Oh Yeah... bloody hell, I'm being yelled at to stand in the corridor outside my room with one sheet in hand. What? Why? It's my father yelling at me! Did I get the wrong message about him being my friend? Apparently, the reason for holding one sheet is so that I'll have to make my bed from scratch before inspection! That'd be right! Now get dressed, make your bed and be on parade in five minutes someone is yelling! Five minutes? But what was it I am supposed to wear? Five minutes?

The morning parade was soon over. Breakfast was a blur. I realised that I must learn to eat faster! After breakfast we were marched away in our companies. I noted that there were a few 'goosesteppers' amongst us, people who walk just fine but find it difficult to march in a squad and swing their arms in unity. They'll learn, I thought. But strangely some of them never did and they ended up as the perennial 'door openers' or seat orderlies or some other non-marching role whenever we had a formal parade. I found myself in 'A' Company.

Off we marched to the soon-to-be very familiar lef... lef... lef-ri-lef:

'In column of threes gentlemen... step it out... swing your arms... HALT!' comes the cry of the orderly in charge of us.

Smash! Several of us civvies crash into the person in front and wonder why they had to stop so quickly? We've never heard any of this stuff before!

We did a lot of marching during those early days. It seemed 'marching us away' was an order that was applied as a solution whenever we were finished with some activity somewhere and no-one quite knew where to go next.

'Form them up, march them away orderly!'

The standard response was... Sir! In spite of most of us being completely clueless as to what was to happen next, someone always seemed to know where to go. It was amazingly comforting to march somewhere in a group, all in step and swaying with a rhythm. A swagger I suppose. I was beginning to understand what it means to march with an Aussie swagger! You don't actually have to think. Just march! Just look at the back of the neck of the cadet in front! Then I was told that every day the staff nominate a duty student who will be responsible for gathering the class, forming it up and 'marching it away'! I panicked at the thought of it. What if I get my lefts and rights mixed up, or worse, I march everybody into a wall or something? Anyway, that'll come. For now, just keep looking at the back of the head of the person in front, and march, in step... lef... lef!

It took several days before we saw our first 'staff' members and the first we encountered weren't the instructors but the Quartermaster's staff. They issued us our gear and there was one hell of a lot of it! Items with funny names like 'pan set messing' or 'knife utility'—it seemed the Army named everything back to front. We were issued greens, the clothes you wear out bush apparently, and a 'greatcoat' with a separate bag of buttons, some of which were big and shiny and made from real brass, and some of which were small, and they all had different insignia on them. The coats smelled of moth balls and looked like they were last packed up straight after the Second World War. These coats seemed to have no purpose at all other than extreme cold weather protection. They reminded me of the coats soldiers always seem to wear in the movies of World War 1. The buttons were confusing. Why not have the buttons on the coats already?

We were given various forms of daily uniform including 'battle dress'—that rough, woolly khaki-coloured uniform that the actors in TV's *Dad's Army* wear—not actually used for battle but as the everyday winter uniform. Next came backpacks and water bottles and sewing kits and boot polishing kits and rifle cleaning kits and some items that I stared at in amazement, wondering how they could possibly be of any use at all! For example, a pristine, green calico bag, shaped into a point, manufactured in Australia in 1943. Someone pointed out that it was a water filter. I wondered if I would ever use it.

Then they issued us our own rifle. It was to be my personal weapon I was told. I'd never even held a gun of any sort before and suddenly I was responsible for one. It had a cold, efficient feel to it. It smelled of steel that had been cleaned and oiled too many times. A gun is a killer. But no, that's not true, the person holding the gun is the killer. The weapon is just the tool. So they told me. But I'd be the person holding the gun, I thought, and suddenly I felt a great weight of responsibility. Was this part of the deal when I joined the Army? I suppose so. I took it back to my room and gingerly placed it on a rack against the wall and stared at it. My 7.62mm Self Loading Rifle, or SLR, gleaming, polished no doubt by generations of cadets who had come and gone before me.

I was also issued a collection of hats. It seemed there was a hat for every occasion: a bush hat, kind of floppy; a slouch hat, as yet 'unbashed' and looking more like something from a western movie; a khaki 'peak' cap for the daily uniform, and a dark blue and red one to go with our 'blues', whatever they were. It was at this point in my life that I realised for the first time that I'm not quite like everybody else. When the quartermaster tried to issue me with my slouch hat, he had to go back to the shelf several times until in the end, he found one big enough.

'You're lucky, that is the biggest size I've got! You have a big head son!'

Backwards and forwards to my room I went to put stuff away and try on my new hats. I couldn't wait to see what I looked and felt like wearing my uniforms. It was one of those things where I had seen many pictures in the advertising brochures of cadets looking so smart in their uniform and I couldn't wait to be one. Within minutes the word had got around how to 'bash' all our hats. It is imperative that you bash the right shape into your hat. Can't be seen looking like some newcomer, must try to look like an old soldier straight away. The officer's peak cap that we were issued had very flat tops which the senior classmen described as 'aircraft carriers', that were impossible to bash into the preferred sloppy shape, where the sides are bent down. A bit like the old Gestapo hats we had all seen in the movies.

Within days, some of the more experienced cadets, those that came from the ranks of serving soldiers rather than the new civvies like me, had organised a group purchase to send away for the caps that we wanted to be wearing as soon as possible. These were the infamous Herbies, based on their brand name of Herbert Johnson, that our senior classmen wore. They were made by the well-known, long-time British military uniform company that makes and sends its hats and caps throughout the British dominions. Would I be in that? Of course. Gotta be part of the mob! How pucka! Couldn't afford to look like a tram conductor in my government issue cap! Had to have a Herbie!

Inevitably, we were marched down to the barber for the short back and sides and then the tailor to be measured for our 'patrol blues', which were not to be worn on patrol but for formal occasions and parades. Another one of those military madness moments when you realise that the name of something is not always an indication if its use or purpose. We were measured for our 'mess kit', which at that stage was just the white jacket and black cummerbund, as we had not yet been allocated to our 'corps'. A 'corps' (pronounced 'core') is a service arm such as Infantry, Artillery, Armour, Transport, Ordnance or Engineers to name a few, each of which has its own mess kit colours. But that was to come later.

Then came the dress lessons. How do you wear all this stuff? When? They reckoned we sometimes had to change several times a day! We learnt to put our uniform on quickly, like really quick, in just seconds! We practised this in a series of what the senior class called 'leaps', which was a drill inflicted by senior classmen upon us inexperienced junior classmen. They said it was necessary because the daily programme forced many uniform changes as we went from classroom lessons to drill lessons to weapons instruction to physical training periods, all in a short time. All for our own good apparently, the leaps, so they assured us, and always accompanied by a good deal of shouting. The leaps consisted of us junior classmen standing outside our doors, in our underpants, and then, when informed of which uniform to put on, racing into our rooms to get dressed, accompanied by helpful instructions delivered in rapid volleys at ear-splitting volume, courtesy of our senior class fathers, who screamed at us which shirt, shorts, underpants and socks go with which trousers, coat, shoes, hats... and in what order. I soon realised that whatever the uniform, it always seemed to have twice as many buttons as anything I had ever worn in civvy street and usually in the most inaccessible places, requiring a deliberate and planned sequence of fastening to avoid discovering that you had left the smallest little button on your fly undone and it would take undoing a dozen buttons to get to it! Who on earth designed these uniforms? Then we would jump back into the corridor in the hope of not being the last one, because not only would that embarrass our senior class father, it also meant twenty push-ups. We practiced leaps in platoon groups until we were all exhausted and the uniforms that were once folded so neatly lay in unrecognisable crumpled piles around our rooms.

The school had an insignia which was a menacing-looking lion standing on a crown inside an oval surround which read 'Officer Cadet School', supported by a banner which read 'Loyalty and Service'. The insignia was represented on all the school paraphernalia, including hat and collar badges, cuff links and name tags. The lions on the collar badges were mirror reversed and worn so that they would look at each other, facing inwards.

One of the cardinal sins was to put your collar badges on the wrong way around, with the lions facing outwards and their bums pointing at each other. This resulted in being yelled at for wearing 'poofter lions'—a serious misdemeanour. More push-ups! Push-ups for an individual were always awarded to the whole section for good measure. All ten of us! I was probably the cause of a large percentage of the section push ups. Naturally, almost immediately after any session of leaps there would be a room inspection. That'd be right! Quick, everything back in order. Remember to make the socks smile and fold the singlets and tee shirts the width of a clue board. I didn't have time to think why all this was necessary. I just did it, to survive!

The days passed quickly, particularly during those first six weeks. We were told that we would not be allowed to leave base or have a drink for six weeks. So, the days consisted of non-stop almost frantic activity, entirely led by our senior class. The staff of OCS, who were collectively referred to as the 'Directing Staff', or DS, comprised the officers and warrant officers who were the managers, administrators and instructors of the school. They gave lectures, ran the drill and weapons classes and managed the other military curriculum, but they generally watched from a distance when it came time to doing the disciplinary stuff. It was the senior class who managed our day-to-day existence and disciplined us on a minute-by-minute basis.

The routine was quickly established: morning parade followed by a quick but huge breakfast; barrack cleaning to include the 'ablutions' (I wondered what on earth ablutions were when I first heard the term); room tidying to include making the beds with 'hospital' corners (another new term); room and clothes inspections; and then form up into 'column of threes' (a confusing term) to march away to the location of the first daily instruction. This would include staff lectures on subjects like government, English expression, military writing, leadership, military history, strategy and tactics and politics. We would be given practical lessons on how to use various equipments such as radios, and naturally there was always drill, every day... endless parade drill where details included such bizarre considerations as the level to which you brought your knee up, and the angle at which you held your lower leg just before slamming it down to stand to attention. Frequently many of us could be seen precariously and hilariously standing for minutes on end with one leg up and one foot dangling, waiting for the 'drill DS' to ensure the manner in which we were about to deliver the foot to ground was the correct one. There were field practices to teach us field skills, such as how to move through the bush in a patrol formation, how to halt and spread out and 'watch your arc', how to set up for a 'night harbour', all the basic skills a soldier needs to survive in the bush. Then there were the many days on the rifle range, learning to aim and shoot:

'Breathe son, relax, bring that sight to bear ever so slowly, now 'squeeeeze' the trigger, gently!'

Everything was referred to as a 'parade': breakfast parade, to eat; dress parade, to inspect uniforms; sick parade, to line up at the RAP (Regimental Aid Post); and sport parade, to go and play not-so voluntary sport. We frequently played sport, all kinds of sport. We would be called out to a 'sport parade' where we would be lined up and trotted off to play sport, a bit like school kids lined up to play whatever sport the teacher thought was a good idea. Being a Queensland kid, the staff reckoned I should be a Rugby League player. I had attended one of the few Queensland schools that specialised in Australian Rules, however, so I lined up with the Rules blokes. Later we even had 'chunder parades'. I am sure these were a form of bastardisation. When the staff knew the cadets had indulged in a night of excessive alcohol consumption, we would be paraded to run off the effects of the night before, often resulting in a few of the cadets bending over double to empty their stomachs!

Interspersed with the daily parades there would always be TOC. TOC meant tea or coffee; always referred to together, never as one or the other, because most times we could not tell the difference. The kitchen staff were not overly careful and frequently mixed up the urns and the coffee or tea brew would be made in an urn that had held the other brew the previous time. A few interesting terms sprang up around TOC. In the evening, we would all dress in our patrol blues, because that is what gentlemen did—and besides, that was the rule. However, running whilst dressed in patrol blues was taboo, so 'TOC racing', which was marching at double time to beat the other company to TOC, became an art. 'Scorched TOC' was the result of losing the TOC race to the other company and meant there was little of anything left.

Then we were introduced to the physical training side of the curriculum. I had been getting ready for this by running five kilometres or so every day at home in Brisbane during the preceding few months. I'm glad I did because it prepared me to run till my lungs threatened to burst: up and down the 'yamas' of the peninsula, as we called the sand dunes, all of which seemed to consist of endless false horizons. One of the yamas was called 'heartbreak hill', a very apt name as the staff always made us do one more run up that hill when we thought we were finished, just to show who was boss. We climbed ropes and dragged ourselves through ditches filled with mud and we carried each other over obstacles and then we ran some more, all in our Army issue gear, which, when running in field gear meant good 'ol 'Boots AB'. We were issued one pair for rough and one pair for parade. Hard as nails they were, made from thick unbending leather, and heavy as lead, relics from World War Two along with the Great Coats. It was a wonder we didn't ruin our feet. The new-issue boots, called Boots GP, which were such a success in Vietnam, had not yet made their way to us lowly cadets.

The daily routine included night lectures and compulsory quiet times in our rooms. Time for writing letters or preparing assignments—a welcome relief with no one to shout at you. The supper bell would ring for TOC and a biscuit and time for bed, glorious bed. We had to practice such essential skills as spit polishing boots, ironing knife-edged creases into our pants and shirts and making the bed with sheets pulled into hospital corners (which probably referred to the field hospitals of the last war, as no modern hospital would insist its staff achieve the precise folds we produced). Whenever we knew there was to be an inspection next morning, we slept on the ground next to the perfectly made bed and hold a fake 'acquired' sheet in our hand at reveille. Old soldier's tricks that us newcomers soon learned! I do recall sleeping on the ground a number of times for no good reason though, because the threatened inspection didn't always happen! Perhaps the old soldiers amongst us got it wrong from time to time. I also figured out that spit polishing boots by the light of a small radiator heater after 'lights out' was not a good idea. It melted the polish that I was trying to harden onto the toecap. Also, don't eat peanuts whilst spit polishing! Seems obvious now but I had to learn the hard way and work twice as hard to remove all the little grooves and divots caused by the partially chewed nuts I managed to spit onto my super shiny boots!

Considering I could barely iron a hanky; never rose without several shakes; and had no idea what the functions of boot polish or washing machines were, I did well just to survive these first few weeks. Had I been a spoilt brat?

A term that became very familiar was 'extra drills' or 'ED' for short. Whenever a junior, or indeed in some cases, a senior classman transgressed in any way at all, he was given one or more EDs. This entailed getting up about one hour before everyone else and, in the pitch black of night well before dawn, getting dressed in full battle gear, which includes back pack filled with all the bush gear, spare greens and hutchie (heavy fly sheet for shelter at night), full water bottles, rifle and webbing with ammunition pouches and spare magazines, and marching off down to where the duty cadet, a senior classman, would be waiting in the centre of the parade ground. After a gear inspection the duty cadet would commence issuing endless streams of parade marching orders... all at double time. This generally resulted in hapless confused cadets marching comically all over the parade ground at double pace, in different directions toting their heavily laden pack and rifle.

This was a sight worthy of any comedy movie, but not funny for the poor cadets on ED who, after a few minutes would be pouring with sweat, even in the cold mid-winter temperatures of southern Victoria. The ED punishment system was an honour system. If awarded an ED, the guilty cadet was expected to go to the cadet orderly room and declare the offense himself by writing it into the punishment book, do the ED, and mark it off when completed. The senior classman who gave the award would check the punishment book to ensure the ED was recorded and completed. The cadet on duty couldn't afford to go easy, in case the duty staff officer, usually one of the Captain Instructors, checked the book and was watching from somewhere in the tree line edging the parade ground. There were many dark ominous spaces around that parade ground in the early hours of the day in which an officer could easily stand undetected. Failure to attend an ED parade, if caught, would end up with more EDs as punishment.

It was not uncommon to see cadets in a never-ending cycle of self-perpetuating EDs. This was because the ED started with a detailed inspection of the offending cadet's gear. If one buckle was out of line, or a water bottle not filled to the exact mark, the poor cadet on ED was awarded another ED. Thus, the same procedure would happen the next morning. If the duty cadet delayed the offending junior classman a few minutes too long, chances were that he would not get back to his 'lines' (an Army term for one's accommodation) on time for the breakfast parade in the right uniform. This risked not only precious minutes to wolf down breakfast, it could result in getting another ED, simply for being late.

At the Officer Cadet School, the world seemed to shrink to the day-to-day activities and the surrounds of the school. There seemed to be no other world. We had no idea what was on the news of the world, let alone back home. That all seemed irrelevant. Just surviving to see another day, in an environment that was designed to severely test us both physically and mentally, was all that mattered. Those that failed the test were removed. I didn't want to be one of them.

3 The Junior Class

The six-week induction was soon over, and it was time to celebrate. Dressed in 'civvies' for the first time, we were herded onto buses which took us into Sorrento, a little local town. Senior and junior class alike, all bundled together, off to have a beer in the Sorrento Hotel. We were not sure how that would go, all in one pub? Maybe they'd spread us out across the town? I wondered how the townspeople would take all this! Over a hundred young men with ridiculously short hair, all dressed the same in grey trousers and blue reefer jackets with Portsea school ties, celebrating with too much beer! Leaning on some bar with too many beers inside me, I remember having a serious discussion with a particular senior classman. I can't remember the detail but the next time I had anything to do with him I was with a classmate called Robyn Patterson and fully laden with our field gear, this same senior classman was running us up and down sand dunes. He was superbly fit and would wait for us at the top of each hill, making us run faster and faster, and each time we made it I would smile at him. Not sure why I did that, perhaps just to antagonise him, show him that I was not broken—an immature nervous reaction to his threats. He became agitated and threatened to punch my lights out if I kept smiling at him. Unfortunately, I stupidly smiled one more time. WHAM! I was on my back spitting out one of my front teeth.

'I told you not to smile', He apologised profusely, as he was helping me up.

We were all big kids on the block, so I smiled at him again with my bloodied mouth, got up and went on running up and down hills. After that smiling became difficult as my mouth felt particularly swollen. Part of Army training is all about stretching your capability, controlling your emotions, not deliberately provoking outcomes but developing calm, reasoned responses to unpredictable situations. He was embarrassed and I had a tooth missing, but we had both learned.

Soon we had our first Sunday off. I sat on the edge of the escarpment with my Queensland classmate Hugh Polson, listening to the Beatles' 'Hey Jude' on the radio, staring out at the free world of the Mornington Peninsula: people swimming, water-skiing, sailing, just having a Sunday off; while we were allowed a few moments of freedom to polish our boots and do 'make and mend'. This was an Army term with which I became very familiar. It was a bit like the order to 'march them away'. When no-one knew what to do with us cadets, they would send us to our rooms to 'make and mend': in other words, go get your stuff organised, do anything but don't bother anyone. Sundays were our only days off, and Hugh and I would spend many Sunday afternoons together, wishing we were with our girlfriends or any other place other than Portsea. Having invested in a toaster, we would scam loaves of bread and packets of butter from the kitchen and eat toasted bread and butter for hours with anyone who cared to venture in our rooms. With the smell of burnt toast wafting down the corridors, there were always volunteers who wanted to join in.

About week twelve, we were given our first complete weekend off where we were allowed to leave the establishment unsupervised! The Queenslanders got together and rented a house somewhere in Melbourne to show those Melbourne chicks a thing or two. It didn't take long to realise that we stood out like sore thumbs! Made to wear the standard reefer jacket, ties and grey flannels and with our haircuts-from-hell at a time when just about every young male was either a surfie or a bodgie, sporting long blonde or greasy slicked hair, we stood out... like... well, like Army cadets! The girls didn't want anything to do with anything Army. I did however manage to meet a pretty girl and invited her to a 'do' at the cadet's mess. She worked for a company that distributed contact adhesive, which is a plastic stick-on covering that we used to waterproof maps. She gave me a roll of the stuff and for a while I was the kingpin of cadets as junior and senior classmen alike begged me for pieces of contact adhesive.

One day during those first few months, the Minister for the Army came for a visit. Naturally we mounted a parade for him: in the Army that's what you do. We had indulged somewhat the night before, and I felt pretty ordinary. Unfortunately, the minister, whilst inspecting the front row, decided he wanted to pause and have a quick chat with me. Me of all people! Perhaps he identified the bloke with the 'hangover' and thought he'd have a laugh! I can't remember what he said, or what I answered, but I do remember having great difficulty in concentrating on his words.

The Honourable Phillip Lynch, Minister for the Army, inspecting the Junior Class, escorted by the Cadet Battalion Sergeant Major John Guy. It was John Guy who had the soothing 'heads up gentlemen' voice that so impressed me on the first morning parade at the college.
Photo: Army Officer Cadet School archives

Soon we went on our first 'bivouac' in the bush, under the stars. The staff marched us away in our platoons and led us to a site where they basically said 'Okay, go to it'. There was no attempt to form us into any defensive positions, no drills or patrols to be organised. This was about us demonstrating our undoubted skills at surviving a night on Mornington Peninsula, which was a cold and windy place at the best of time. I had been on a few YMCA camps in the bush before but that was more about sitting around a campfire, toasting marshmallows and singing 'Kum Ba Ya'. I stared at the pile of equipment that rolled out of my pack: a large plastic sheet that someone called a 'hutchie'; some long plastic tubes, called 'inners', that we were supposed to inflate with our mouths and insert into the rolls of a groundsheet called 'outers' to make a mattress; some bits of green string called hutchie string and two old army blankets with press studs that seemed to come from the same stock as the smelly great coats and the battle dress and the green calico bag. Then someone was yelling to come and get our food and we lined up with our aluminium mess tins. Or, in that Army back-to-front speak, our 'pan sets messing'. Lined up on the ground before us in the field kitchen were weird looking Army green-coloured 'eskies', or what we later came to identify as the ubiquitous 'hot boxes'. These had probably not changed since WWII and by the time I left the Army some 24 years later, this same style of hot box was still in use!

What a sight we were. All these blokes who had rarely, if ever, slept under the stars, making a racket stuffing our inners into our outers with a balloon-squeaking noise that carried for miles. Whoever invented these inners and outers clearly misunderstood what is involved in evading enemy forces. None of us ever used the inners again! We erected our hutchies against the rain, rolled the blankets into a make shift sleeping bag, and fell asleep on a cold miserable night, gear scattered everywhere. This didn't seem very military to me, and later I realised that this was the lesson in how not to do it, as very soon afterwards the DS proceeded to teach us all about tactical sleeping in the bush. To this day, whenever I am camping, I put my stuff exactly where I know it will be if I have to get up in a hurry in the dark of night.

A brew in the field, a rare moment of relaxation whilst learning bush skills at Puckapunyal Training Area. Note the 'foul ground' sign just over my shoulder.
Photo: Hugh Polson

After that first night out under the stars we came back to our lines (barracks) for the complete opposite. Instead of dining in the bush, eating out of hot boxers, we changed for the first time into our mess kit and had our first 'dining in' night, those traditional, very formal dinners that the military is very good at. Candelabra, lines of silver cutlery and rows of wine and port glasses. We toasted the Queen. We felt a million dollars!

We learned much that first term. We learned to respect the each other and the value of mateship. We respected the instructors, although we had our reservations about some some. The 'PTI's', our physical training instructors, whose voices would send us over the sand dunes and under the barrels into the mud and up the ropes and running in squads till our lungs burst, we truly hated them at the time, but in our hearts, we knew the truth. They were tough men and what they really taught us was:

'Nil Desperadum—Never give up'.

'Never give up son, even if you think you have no more to give. There is no limit to what you can do. Don't be selfish. Look after your mate because in time of need he will look after you!' I can see their faces now.

We bonded like young men who were determined not to be broken. On one particular nine-mile forced 'run-march', dressed in all our field gear and carrying rifles, we had to march-jog-run hard, in order to make the allotted pass time. Some of the cadets went down with fatigue. But instead of leaving them behind, a fellow cadet would take their rifle and another their pack and another would give a shoulder, so that the exhausted cadet would have time to recover and once again join in. Just as we were approaching the finish line on that nine-mile run, a Malaysian cadet called Arafin, who was running next to me, collapsed with little warning. His head started to wobble so two of us on either side grabbed him under the armpits, lifted him bodily and carried him the last twenty or so metres to the end, where he literally floated across the line. Luckily Arafin was not a heavy chappy, being of small Asian build. He was very thankful, because anyone not making it to the end would have had to do it again, this time without the benefit of running in a group, which is always a much harder task.

It was in those first few months that I learned about Army 'Padres', and the 'Sally Man'. The Padre was the euphemism for an Army Chaplain and the Sally Man was a collective description for the wonderful Salvation Army men and women who would brave the cold and always be there with a hot brew and a biscuit, and in those days... a smoke! The Padres would provide a cheerful comment, and a smile, and they seemed to know instinctively which of the cadets was feeling vulnerable, take them aside and give them encouragement. They would turn up just when you needed them including, as I was to learn later, in Vietnam! Even in a theatre of war! I vividly recall a particular Padre, whose religion I cannot remember—and quite frankly their religion was never an issue—sitting without a weapon on an Armoured Personnel Carrier, heading out with an infantry battalion on patrol in the Vietnamese jungle. He just wanted to see what 'his boys' went through. He was a good man. I had many an animated discussion with him and never about religion.

A strange training event near the end of those first six months, just before our senior class graduated, was called 'Revenge'. I'm not sure why it was called by that name, as it wasn't not us junior classmen that were able to take revenge on the senior class. Instead, it was they who were able to extract their last bit of meanness on us. We were bussed to the Victorian mountains at Taggerty near Toolangi. Beautiful bushland with tall Mountain Ash eucalypts, the area is rugged, isolated and perfect for an activity designed to be a bit like 'search and evade' training. The junior class was broken into small sections of six cadets, each with a member of the DS attached. Each group was given a rudimentary map, a compass, a 100kg bag of wood to carry, and no food! In fact, we were all searched prior to leaving to ensure we had no goodies stashed. The idea was to find checkpoints along a 50 kilometre route in dense, rugged terrain while carrying the bag of wood to a final rendezvous. Some of the checkpoints had food available. The surprise factor in all this was that the senior class were out there. We were supposed to evade them, but they were out to find us and when they did, they were allowed to inflict punishment. This punishment might include taking away a cadet's bootlaces, or if a group was caught a second time, their boots, or some other item of clothing. We were told the senior class would patrol and guard the roads, so the idea was to keep to the hills, stay away from the obvious routes, and bush bash. This made it harder to carry the 100kg bag, up and over the hills and through the dense undergrowth.

Our group did well. At the first checkpoint, we found the Sally Man who gave us a hot drink each and a tin of bully beef to share! On the second night, the staff at another checkpoint gave us live chooks... one each! A live chook? What were we supposed to do with a live chook? Luckily, they also gave us some matches so we were able to cook the chooks on an open fire. It snowed that night! Lighting a fire and cooking a chook in the snow has its moments. Just killing the chook humanely was hard enough! The next night we deliberately avoided calling in at the checkpoint because our group hadn't been caught yet and we figured they'd all be out to get us,

A group of hapless Junior Class cadets during 'Revenge', having been given live chooks to eat after two days of no food at all. The task of killing the chooks, preparing them and cooking them in the midst of a snow storm was a challenge.
Photo: Hugh Polson

somewhere in the approaches to that last checkpoint. They never found us, or at least we thought they didn't. We did find a sock, tied to a tree, which had an Army ration-pack tin of Vienna Sausages in it. We couldn't understand how this came about. Had the senior class been watching us quietly and were they happy with our approach, thus rewarding us? As usual our DS, who was silently shadowing us the whole four days, said nothing. But I did detect a slight smirk when we found the sock. Revenge has its mysteries.

So, my group made it, one of the few groups who had managed to avoid 'punishment'. After four days of hard slogging, they lined us up and marched us away, again: a 20-kilometre forced march. We were relieved to find the buses lined up when we reached the end and, even better than that, we spotted lines of steak and sausage-filled barbeques. Every senior classman, every staff member and even all the Sally Men (and women) were there, cooking steaks and sausages and rissoles by the hundred, all for us very, very hungry junior cadets.

4 THE SENIOR CLASS

The coming of Christmas 1968 signalled the end of the first six months and the end of our time as junior class. We were about to become senior class to a bunch of unsuspecting civvies. The school closed for the Christmas break. Four glorious weeks off! The Kenyan cadet, Silas Seet, had nowhere to go so I invited him home to Queensland for a few days. That led to a memorable evening. Lo was in Sydney for Christmas with her family but Silas and I were in Brisbane. We wanted to go out on the town and have some fun so I arranged a date with a girl that I used to see at Sunday church (my parents dragged me along), way back when I was in high school. I explained I had a mate up from Melbourne and could we make it a double date? I offered Silas as her friend's blind date, but gave nothing away about his background.

Picture a tall, well-muscled, very black man with a huge, wide-mouthed smile that would disarm anyone even remotely reticent. That was Silas. When we picked up the girls, their world momentarily stopped. Stunned disbelief, followed by genuine surprise, a giggle and 'oh shit what are we doing?' But Silas turned on the charm until they relented and we found a live music pub in Brisbane where we sat right next to the dance floor and drank and laughed. That night, the pub patrons of conservative Brisbane of the 60s would be unlikely to forget this black man doing his 'warrior' dance. Thirty years later I ran into the one of the girls, and she remembered Silas well. Who Wouldn't?

Once back at Portsea we, as the new senior class, planned our reception for the hapless bunch of young men that would soon make up the new junior class. Some of us had been given 'cadet' rank. I was now Cadet Corporal. This meant that I would be the father of two junior classmen instead of one and was responsible for a 'section' of ten cadets. One of our classmates Sam O'Shea, was appointed the BSM for that term, and he decided we should all get extra short haircuts, even shorter than the standard short haircuts. The on-base barber removed the slightest hint of any hairs on our heads more than a millimetre long and we were ready.

The next day the clueless newcomers arrived. We knew the drill. We met them with an air of superiority, a silence that was deafening, disdain for their lack of knowledge about everything. We strutted about and did the 'look tough' bit, and made them wish they had never left their families behind. We could sense their insecurity. We fed on it, temporarily but deliberately destroying their self-confidence so that we could offer them tit-bits of sympathy from time to time. We made them look for solace in each other to begin that build-up of mateship. It promotes the 'let's stick together so that these bastards can't get us' syndrome, a technique well employed by the military. We treated them to leaps and the many other acts of bastardisation we had ourselves been subjected to. I decided my junior classmen should clear a minefield that had suddenly sprung up in the parade ground. They were to use only toothpicks held between their clenched teeth. Juvenile? Yes. Fun to watch? Yes. Did it achieve anything? Probably not.

As senior class we were now responsible for the discipline of the junior class, but we had our own pressures. There were frequent examinations in subjects like military history, military strategy, government, English expression as well as daily skills tests for field subjects to include radio communications, drill and weapon lessons, and marksmanship. We had to be competent users of various weapons, including the 7.62mm SLR, 9mm pistol and the 9mm F1 sub-machine gun.

Tactics: in this case the hands-on actions of a Platoon Commander during an attack or defence or other action, was a stressful subject to us non-Army background senior classmen. The tactics testing events were called TEWTs: 'tactical exercises without troops', where cadets were tested on their knowledge of tactics by having to analyse complicated fighting scenarios that comprised virtual deployment situations of your own platoon plus other units and formations, both friendly and enemy. Offensive and defensive military equipments, such as artillery guns and tanks, were set in virtual locations on the ground, or 'in the field' as the lingo went. It was a hilarious sight, seeing dozens of cadets running around the hills and valleys deploying 'pretend' Army units in positions, by bashing coloured pegs into the ground. That was the easy part. The hard part was to explain to the DS, why you did what you did. For example why you planned to attack from the left when it was blindingly obvious that you should have attacked from the right. I did just such a thing on one of our final TEWTs, muttering my reasons in a less than confident manner, wondering why I had been so stupid. Clearly all the other cadets had wacked in their pegs, denoting their attack start line, on the opposite side. I dreaded the result, expecting a failure, and when the class result list was posted, looked for my name somewhere towards the bottom of the list. Panic, it wasn't there. I must have failed! But wait... there it is. I actually achieved a top three result!

'Well done mister' it was the DS, the one to whom I explained my attack. I wanted to say it was luck but held my tongue and accepted the praise. It was short lived. The next TEWT I barely passed. Oh well.

I had decided to take my car back to Portsea with me for this second half of the course as, being in the senior class meant we would have more spare weekends and it would be good to have some wheels to get around. Being a closed camp, Portsea always had guards on the gate, which made it difficult to get in or out without a pass. I was keen to go and visit Lo, who lived in the hills district just north of Sydney. From Portsea, this was every bit a 12-hour drive, and as we were never given leave passes that were any longer than 48 hours, a trip to see her was always going to be a substantial effort. But, being young, spirited and full of ourselves, four of us from the school decided to make the trip when the first 48-hour pass came up. Hugh Polson, my friend from the start, was keen to come, and his 'son', Kim Duffy, also had a connection in Sydney who he wanted to see. The fourth member was Arafin, the Malaysian cadet.

I hatched a detailed plan to ensure that we would get the maximum time away, and cleverly arranged for the local garage to come and pick up my car for a service. The mechanic was complicit in the 'crime'. He knew to leave the car at the garage (therefore outside of the college walls) after the service, with the keys hidden in an agreed place. That way there would be no need to drive through a guarded gate, before our leave commenced. The four of us would escape on foot, avoid the main gate, walk to the garage, find the key and get away to Sydney, a full twelve hours before official leave actually started. The escape route was to be via the cliffs and beaches on the Port Phillip Bay side of the school. Being on a peninsula, the school has ocean beaches that face Bass Strait and tall cliffs dispersed with calm beaches that face the bay, accessible only when the tide is out. This not-so minor detail of tide times had not entered my planning, but we were lucky! On that Friday night the tide was out, albeit on the way in. It left us just enough room to drag our gear over the rocks and boulders and bits of barbed wire fencing, that probably dated back to the days the place was a quarantine station.

Feeling upbeat we walked the three kilometres to the garage; any minute now we would be free and tearing up the highway. But then, when we got there... where was the car? Incredibly, one of the mechanics, obviously not one that I had spoken to, had decided to do me a favour and actually return the car to the Portsea carpark.

'Nooooh!' What a disaster!

I copped some unfair abuse from the other three but there was no way out if this. We would have to backtrack, get the car out of the carpark, and do a daring drive through the front gate tonight. Or, we could wait till morning! But the tide? Ah no, now we were in a panic. Would the water level let us back in? Luckily it did, and quickly we scrambled back over the rocks, just making it with not much more damage than wet shoes and socks. We retrieved the car and did what we could have done from the start. We drove straight to the guard on the gate and appealed to him to let us escape early. Strange looks from the guard confused us. He couldn't understand our nervousness. By now it was after midnight and he pointed out that the weekend leave started at midnight! Bloody hell! All that for nothing. Oh well, we were on our way!

After a memorable time away in Sydney, during which I thought it was perfectly reasonable to crash on Lo's parents with my cadet friends they had never met, we made another not-so-wise decision. We decided to use the Princes Highway for the return journey to Portsea. This road follows the coast and takes in all the little coastal towns. Great thought, but was it a good idea? We had already left our departure from Sydney to the very last minute, and instead of the expected 12 hours, it took us over 14 hours to complete the trip back. We were supposed to be getting ready for a week-long field test exercise at the school... taking the long road home was clearly a dumb decision! The exercise was the major and final 'leadership' test event for the senior class, during which the DS would randomly appoint individuals to assume senior roles in a platoon, including platoon commander, and asses their management and leadership performances.

We scrambled to get back to Portsea in time to pack our bush gear and just made it to the bus for the 0530 departure. The four of us looked washed out, as if we had been on a binge, having had little sleep in three days. We worried that this would draw the staff's attention, which would not be good and yep, that's what happened. Not having had time for a shave, which at the cadet school was akin to a mortal sin, I stepped off the bus hoping to meld into the background and let some other cadet face the music of being the first one appointed to be the platoon commander. I just wanted to bum around in the background as a riflemen or something and sleep! But, I can still recall that calm, authoritarian voice, from one of the DS captains:

'Ahhh... Mr Lans, what have we got here? No shave, huh? You're a grub mister! You should be ashamed! Step over here, I'm going to make you platoon commander for the first day!'

My worst nightmare! This isn't happening. What does a Platoon Commander do again? I remember the smirk on the DS's face as he watched me panic. I gave some orders that sent people scurrying off into the bush and I sat on my pack to think about things awhile. However, in the end, I unscrambled my brain and I think it went okay. I graduated, so my performance must have been at least acceptable.

As the year drew to a climax, the pressure mounted. It became obvious that some of the cadets were on the edge and struggling to make graduation. Being given a 'free ticket to Frankston' was the term used when a cadet was dismissed from the school for underperformance. There hadn't been many underperformers during our year, and any dismissal usually happened very quickly, with little fuss. I remember well, the line-up of cadets outside of the Adjutant's office, all with very sombre looks on their faces, receiving the 'shape up or ship out warning'. Fortunately, I was never on that line.

At last, the countdown to graduation came. Lo was going to be there, along with my parents and brother and sisters and many of their family members. The day arrived and Hugh and I drove to the other side of Melbourne to rendezvous with my family, who had driven from Queensland and Sydney, and escort them to their apartments in Sorrento, not far from Portsea. Later, I went to pick up Lo who arrived on the train. That was a special moment. For much of the year at Portsea I had been imagining this time, when graduation would come and Lo would be there.

The day of the parade was sunny, with mid-winter blue skies, but it was intensely cold. Our families suffered in silence. As a class, we were well versed at doing parades, only six months ago we were on our senior class graduation parade. There were no hiccups for us and at the end we threw our hats into the air and cheered! The graduation ball was held that night. Here we were, dressed to the nines in our regimental mess kit. Very colourful, with each corps having a different combination of coloured coats, lapels and trouser stripe. I was allocated, as requested, to the corps of Artillery. Artillery was my choice because I wanted to be in an 'arms' corps (a fighting corps instead of a support corps) but primarily because the School of Artillery was at North Head in Sydney, where I would be doing further training for the next six months. Lo lived just outside of Sydney. What better reason than to choose Artillery?

The ball was held in the cadet's dining room with dancing in the adjoining ante room. This area was once the dining/ante room for the quarantine station which occupied the site in the previous century. Must have been a lot of people in quarantine in those days as the area was well capable of holding a couple of hundred people. A live orchestra made for a great night and Lo and I danced into the early hours, part of a noisy and happy celebration. Then came the moment that we were waiting for, the ceremonial 'pinning of the pips'. Each of the graduating cadets kneeled down to have their badges of rank pinned on their shoulder lapels. One solitary officer's 'pip' on each side, signifying that we were now 'Second Lieutenants'. Most of us chose our mum to pin one side, and our girlfriend or wife the other. And that was it! When we stood up we were no longer officer cadets. We were officers! It reminded me of a cartoon that was being flashed about the place, showing a starry eyed Second Lieutenant marching proudly away above a caption that read: 'twelve months ago I couldn't even spell *occifer* and now I *are* one!'

The next morning I said goodbye to the place and the people that had been so much a part of every minute of my existence for a year. This was where I had wanted to come, and almost immediately upon arrival, wanted to leave! It was where I grew up; where I went from being a boy to a man. I was a 21-year-old spoilt youth when I arrived, and a 22-year-old adult who left. At the Officer Cadet School where many challenges were dished up to me, I learned to stand up and be counted as part of the team... the Army team. I learnt that, no matter how tough it gets, things can always get tougher, and there is no time to waste on self-pity, because someone may well be relying on me to make a decision that may affect their life. This place, these instructors, these new Army mates, would remain a significant part of my life.

Portsea changed everything. It was here that I learned when to lead and when to follow; where I learned to accept and discharge responsibility. It was here where I learned what it is like to rely on others and have them rely on you.

I went back to my mates in Brisbane and initially felt a little distant from them. I had changed. Was I suddenly so mature? Not really, but I was developing and laying new foundations for my future. I admit there were times in the last twelve months that I wondered if I had done the right thing. But now I was sure. A career in the Army was what I wanted. Bring it on!

5 AUSTRALIANS IN VIETNAM

The Australians in Vietnam were there at the request of the Americans. Australia, like America, subscribed to the 'domino theory'. If Vietnam was allowed to turn communist, what country would be next? Malaysia or Indonesia? The Australian Army, organised into the 1st Australian Task Force (1 ATF) and the Australian Logistic Support Group (ALSG), was responsible for the security of Phuoc Tuy Province. A condition of the Australian government for Australia's entry into the war, was to allow Australia to manage the conduct of their own counter-insurgency measures in a province of their own. This would ensure that Australia could develop and practice Australian counter-terrorist tactics and be responsible, and answerable, for their own successes and failures. Thus, the Australians were allocated Phuoc Tuy Province.

Phuoc Tuy, approximately 100 Km north-east of Saigon, is approximately 100km x 120km in area, with some large hills near the coast that were referred to as the Long Hai Hills, the Nui Thi Vai Hills and the Nui Dinh Hills. The main access road, Route 2, runs straight up the middle from south to north, servicing the rubber plantations and a few hamlets. Vung Tau in the south was the main logistical base and the tactical positions occupied by the Australians were Nui Dat, on Route 2, smack in the middle of the Province, and a permanent Fire Support Base (FSB) called 'The Horseshoe', where an

Australian gun battery was located. Vung Tau was the largest town and one of the major ports in the southern parts of the country. Vung Tau, in addition to being a logistical base, was also a place where Australian soldiers could have a short period of leave, as a break from the operational environment of Nui Dat and the rest of the Province.

The 1st Australian Task Force comprised two, or for a period of time three, infantry battalions with supporting artillery, armoured, engineer, intelligence, signals, transport, aviation and special forces as well as medical, Air Force and logistic support elements: a total force of around 5000.

I had arrived in South Vietnam. This was a country at war.

'It's hot... man it's hot! Excuse me, could you tell me where the Australians are?'

It was January 7, 1970. I was standing on the blistering hot tarmac at Tan Son Nhut airport in Saigon, wearing my Australian 'polys'. The Army's day uniform was referred to as 'polys' because they comprised long trousers and a short-sleeved shirt, made primarily of polyester, with a small percentage of cotton; entirely unsuitable to my hot, sweating body which hates anything other than cotton! Held in my left hand was a large duffel bag with the tri-coloured kangaroo and '1 ATF' stamped on it.

It seemed a ridiculous question, but a group of us had just disembarked the Defence chartered QANTAS 707 that had delivered us from Sydney to Saigon, and we were standing on this tarmac like lost sheep, waiting for someone to round us up. We were 'reos', short for re-enforcements. The young Australians with whom I was standing were all there to replace repatriated Australian soldiers who had succumbed to illness, such as malaria, been wounded or killed. However, when we arrived at the airport, nobody seemed to know who we were, or where we should go.

In those days, Tan Son Nhut was reputed to be the world's busiest airport, used as the primary point of transit for the half million or so American soldiers that were there, plus all the other contingents from the allied countries supporting the US effort in Vietnam.

'Okay you Aussies, just follow me,' came a comforting command at last.

Soon we saw the more familiar surrounds of the Australian Army's Movement Group—Saigon. By now I was a wet, sweaty mess, trying to deal with the debilitating humidity and the phenomenal heat reflecting of the melting tarmac. They hadn't told us about this at Canungra! Profusely dripping in our 'good gear', we were ushered onto a RAAF Caribou by a relaxed Aussie Sergeant wearing only shorts and no shirt, his rank evident on a band on his left wrist.

We were all familiar with Caribous, the only aircraft in the world that flies so slow it appears to be going backwards when flying into a headwind. That was popular belief anyway. The Aussie pilots soon had us landing on Luscombe Airfield at the 1 ATF base at Nui Dat, in Phuoc Tuy Province. From there the group separated, met by their various unit representatives, and before long I found myself in a Land Rover, winding its way through tented regimental areas scattered amongst rows and rows of plantation rubber trees. It came as a surprise to see just how big this area was. I was experiencing an overwhelming sense of claustrophobia, with intense heat, humidity, dust and mud, more or less all at the same time pressing on my senses. Wow, what a climate! How was I going to survive this place for a year?

Soon I was standing in front of the Adjutant of 1 Field Regiment, but instead of a warm welcome, it was:

'Don't put you bags down Lans, you're going forward.' The comment was delivered by Captain Don Banks, with a smirk which said: 'Suffer, newcomer, you are about to learn what it's like to deploy in a war zone!'

'The Commanding Officer wants you to go forward as a Section Commander for 101 Field Battery. They're out on operations occupying Fire Support Base (FSB) Peggy, which is up Route Two about an hour away. But don't worry, I've arranged a ride for you on the Chinook re-supply helicopter this afternoon.'

I had arrived at the same time as the other two 'graduates' of the same artillery course back in Sydney. One of those, Geoff Ryder, was becoming a good friend and we were almost clinging to each other like long lost buddies in this strange new world. However, he was whisked away to his battery at The Horseshoe. I later learned that The Horseshoe had played a vital role in the successful defence of Nui Dat during the battle of Long Tan, two years before, where a Kiwi artillery forward observer officer had played a major part in saving an Australian infantry company that had intercepted two regiments of North Vietnamese Regular Army on their way to Nui Dat.

I stared at the Adjutant. Okay, so this is getting ridiculous. I'm still wearing polys! I'll be stepping out on active duty wearing Sydney clothes! But then he added:

'Drop your bags here and get changed into some greens, go to the Q Store and pick up your webbing and rifle and ammunition and your pack. Anything that you forget can come out later.'

Sure enough, within an hour I was on the move again, looking down at the battle-scarred countryside of Vietnam with my worldly belongings distributed between my backpack and webbing. On the webbing hung a 9mm Browning service pistol with ammunition magazines, and across my lap was my 5.57mm US-made M16 semi-automatic rifle. The Australians used 7.62mm Self Loading Rifles (SLR) in Vietnam, which were a much heavier and more capable weapon than the M16, with a great deal more 'stopping' power. However, as an officer I didn't need to carry an SLR. The M16 and the pistol were sufficient security. I needed to be more hands-free to read maps, answer radios and look through binoculars and hold compasses, clearly there were not enough hands to do all those things simultaneously. This was for real. Until now I'd only handled live ammunition at a firing range. Carrying live ammunition around with me for every minute of the day, in the same manner as one might carry a wallet or a watch, seemed strange. It wouldn't be long before I would feel naked without my weapon at my side, always within arm's reach.

Flying out to FSB Peggy, it quickly became obvious that I was now in a country at war. Everywhere were the scars of battle and destruction:

barricaded villages with tangled barbed-wire surrounds; burnt-out, wasted patches of ground laid bare by artillery barrages; and large craters where 500lb and 1000lb bombs, delivered by B52s, had created deep holes, large enough to permanently fill with water, giving them the incongruous appearance of swimming pools. The jungle was everywhere, deep and, from the air at least, impenetrable, except in patches where it had been felled and replaced by rows of rubber trees. We were following National Route Two directly north, but I was without a map, so I had little idea of where I was in relation to Nui Dat. In Vietnam, there was no front line. This was guerrilla or 'counter-insurgency' warfare, as I was taught at officer school. This was not *conventional* war, this was *unconventional* war and we Aussies were supposed to be very good at it. The enemy was everywhere and anywhere. The Vietnamese all around you were both friend and foe; either way, they looked or dressed no different. I learned later that the enemy included the 'papasans' and 'mamasans'. Even the brother or the sister of the man in the village, or the son or daughter of the peasant who tilled the land, might step out into the jungle at night and come to kill you. And the next morning, they would return to be the innocent villager again. The enemy was not just out there, he or she was all around us in the villages, on the roads and in the crowds, in those eyes that followed you as you drove by.

Vietnam is a beautiful country. It was even then, despite the war. The French had long been the colonial masters here, until their humiliating defeat at Dien Bien Phu in 1954, and had left behind many a beautiful house in country estates that had flourished, growing rubber or palm oil. Deserted seaside villas with crumbling grand entrances, accessed through long driveways lined with poinsettias and frangipanis, were a feature of the landscape in Phuoc Tuy. They were once plantation managers' holiday residences. The style and character of French colonial architecture contributed to a distinct French Quarter look in many regional towns.

Soon enough we landed at FSB Peggy and I stepped out of the Chinook, finding my way through the piles of supplies and hot boxes full of fresh food for the soldiers manning the base. An FSB, I quickly learned, was a secure base situated to support an infantry operation. FSB Peggy consisted of an artillery battery of six 105mm artillery howitzers (referred to by us simply as guns), armoured support comprising three Centurion tanks and a troop of four personnel carriers. In addition there was a company of 85mm mortars, all surrounded by 'bunded' earth and machine gun posts. Earth that is bunded is earth that is pushed up into protective barriers.

Fire Support Base Pamela (the 105mm howitzers have just fired), constructed and occupied by 'A' Battery, 12 Field Regiment in June 1971. It was similar to FSB Peggy and typical of Australian FSBs. Note the earthen bunds pushed up to protect the gunners from ground and mortar fire. The enemy had no air support so camouflage from the air was unnecessary. Individual sleeping bays were dug into the bunds.
Photo Commonwealth of Australia DVA

The Chinook landing was another first. I experienced the incredible chaos created by the large twin blades of a Chinook. Surely, they create the biggest dust cloud of any helicopter! It blows dust into every orifice of your body, it seems. Eventually the dust cleared as the chopper noisily beat a retreat, and I looked around in the vain hope of finding a friendly face. That's when I heard the words:

'G'daaaay, Sir... the Gun Position Officer has asked me to look after you! Just come this way.'

There stood a massive man, hairy and bare chested, rifle in one hand, dog tags dangling around his bull-like neck, and a piece of green cloth around his wrist which had three stripes sewn on it. His name was Buck and he was a Gun Sergeant. Buck had been in Vietnam for over nine months. In Australia, Buck would have looked like he should be riding a Harley Davidson motorbike with his bikie gang. Here in Vietnam he looked exactly like the type you would choose to protect you against the bad guys out there beyond the wire. Sergeant Buck was well known. A trustworthy NCO who ran a tight gun crew and the right person to introduce a brand new, young and inexperienced officer into the ways of the Australian gunners in Vietnam.

The next level of my Army education was about to begin. The new officer in Vietnam, still in nappies! This was an important step as a young officer. This was not about Vietnam. This was not cadet school or an artillery course. This was for real, the here and now. The men I was about to meet would study my actions, weigh up my responses and judge me. These next few days and weeks and months I would begin my career by learning from the Senior NCOs with their wealth of experience. The military system may seem confusing because of the rank structure, but it works well. Soldiers join as recruits, promoted to reward excellence, amassing practical experience as they progress up the ranks so that by the time they reach senior NCO ranks, such as Sergeant or Warrant Officer, they are competent experts in their field. Commissioned officers join as young cadets and receive training that lays the foundation for command and staff positions. A young Second Lieutenant on graduation is 'senior' in rank to any of the NCOs, through formal education and training but has little or no experience. Young officers need to understand that when they start out, they must learn from the NCOs that they 'command'. Here I was in Vietnam at the start of my career as an officer and, looking at Buck, I was happy to start by listening and learning from him.

My mind went back to only a couple of days before when I was sitting with Lo in the back seat of my father's car, heading for the Eastern Command Personnel Depot at Watson's Bay, Sydney, from where I would be taken by bus to the airport. We were oblivious to anything or anyone around us. My parents were in the front seat and my mum was beside herself with worry and frustration. Only a few years before, in 1966, she was unaware of the Vietnam war. The protests had not started—not seriously anyway. I was just a young fellow working in a bank and going surfing on the weekends. One memorable day I was sitting with my parents on the front veranda of our home at Aspley in Brisbane, when a bare gun carriage slowly rolled past, towed by an old fashioned looking Army truck. It was a funeral. This was not unusual in itself, as our house was on the road to the cemetery, but this was not just any funeral. On the carriage was a coffin, draped in an Australian flag, and on the flag was fixed an officers peak cap. My parents stared, a shocked, silent and painful stare. Here was one of the first Australian casualties of the war in Vietnam. My mum blurted out that thankfully her family would not be involved in it.

She had given much, my mum. She was the matriarch, fighting for her family's safety, right throughout the World War II Japanese occupation of Indonesia. She intervened on behalf of my brother Reg when he was in a Japanese goal, sentenced to life imprisonment for being part of the underground opposition. She shielded and nurtured my two sisters during the occupation and when they were put into a Javanese women's camp. When the war ended, my father, who had been in a Japanese prisoner of war camp for three years, managed to re-unite the family over a period of nearly two years and took them to Holland where they were safe. In Holland, my mother wanted to go to this wonderful new country called Australia. Her part of Indonesia was liberated by the Australians and she identified with the young Aussie diggers who she said were honest and trustworthy, larrikin-like, who made smart remarks and loved her children. Yet here she was, over twenty years later, watching as another war, albeit nothing like WWII, was slowly but inexorably sucking into it young men from her adopted new country, and now killing them, just like the one that she had seen on the gun carriage. Just four years after being shocked into realising that Australian men were dying in South Vietnam, she was putting me on a plane to that very place, to go to another war, one that she could barely understand.

My journey from Portsea to Vietnam was quick, and deliberate. I was allocated to the corps of Artillery upon graduation, which was my first choice, partly to be close to Lo, who at that time was at Sydney University doing a degree in education. The School of Artillery at North Head was in a glorious location, with views overlooking Manly, North and South Head and Sydney Harbour, right up to the Harbour Bridge. There were about thirty or so young officers on my course. We Portsea graduates were joined by the 'Officer Training Unit' (OTU) graduates from Scheyville, west of Sydney, where selected National Servicemen, or 'Natios', were trained to become officers.

Having just survived officer training, we were suddenly officers, who really didn't know how to deal with that new-found status. We had graduated into an Army still bound by quaint old customs, many of which would soon disappear. One of the strangest of these was the Army's penchant for 'batmen'. A batman looks after *his* officer. This goes back centuries, where officers in the British Army would have batmen to look after their comfort and carry their extra gear, iron their uniforms and generally be ready at their call. As time went on, the role of these batmen changed. In the Australian Army, officers of Major rank and above in regimental postings had batmen on their staff. However, the role of the batman had evolved to become a 'jack of all trades' who was there to look after their officer when he did not have the time to look after himself. The officer might be planning missions, writing deployment orders, directing artillery fire or doing some other 'officerly' thing.

However, at the School of Artillery, the Officers Mess had a permanent allocation of batmen, ex-soldiers, employed part time to look after the young officers. We were told to leave our boots and shoes outside our doors for the batman to polish. I felt embarrassed! These batmen were old blokes, who would wake us in the morning with a cup of tea. Delicious it was too, best tea ever! They made our beds and cleaned our rooms. One of the batmen had a strange habit of placing a full cup of hot and sugary tea on our chests saying 'come now young sir, it's time to get up!' You can imagine the consequences for the young officer not quite awake: a very sudden and sodden experience!

Upon arrival at the artillery school in July 1969, we were welcomed by the Chief Instructor with words that are still as clear now as they were then.

'Gentlemen, the 4th Field Regiment needs an additional four officers when it deploys to Vietnam early next year and it's quite simple: the first four placegetters on this course will be posted to 4 Field and go to Vietnam with that regiment!'

My mind was made up. Graduate in the top four come hell or high water. I did! I came fourth. Three Natios beat me to the first three places but I was the highest qualifier of the regular officers. When I proudly, but stupidly without any sensitivity, told my mum that I was off to Vietnam, she turned white. She took it badly. I should have known. I thought of the coffin on the gun carriage and her reaction and here she was having to send a son to a war in a country that she had barely heard of, for a cause that most Australians were not even sure of. I, on the other hand, had been imbued with a sense of righteousness. To us young graduates, intervention was justified. We were off to stop the 'yellow hordes', or something like that. In a democratic society like ours, the Army does what it's told by its political masters, and our political masters told us to go and fight for freedom in this hot, faraway place. We were going to save the 'free' South Vietnamese from being overrun by the 'communist' North Vietnamese and stop them from marching southward, taking country after country. It was called the domino theory. This was my job. I had trained for it. How naive I was!

6 Going to the War in South Vietnam

I barely slept on my first night in Vietnam. I was lying under a smelly mosquito net on top of a bed of 105mm ammunition containers, in a hole in the ground with sandbags all around and a thick layer of dust on everything, my rifle at hand and my boots nearby ready to pull on. My mind raced. Why did I want to come here? How did I get here? I remembered:

Somebody is yelling at me. It's dark though. Why so early? Ah, that's right. I'm at JTC (Jungle Training Centre) in Canungra, in bloody hot Queensland. I'm lying on a canvass stretcher and there are Army blokes all around me. I get up quickly. They told us yesterday, when we arrived, that there's no respect for rank here. We're all the same. Army or Air Force, Corporal or Colonel... we all have to do this bloody three-week Vietnam toughening-up course. It's the rule! As I line up with the others, I look around at this motley bunch. All sorts really. Blokes from different corps and different units. Reos like me.

The NCO instructors are standing to the front, hands on hips, inspecting us casually. Then someone speaks: a red-headed Sergeant, a young bloke with a friendly face. Another one of those confident young men. He's been there; he knows what it's about. Next to me is my new mate Geoff Ryder, a Scheyville graduate. He's a Natio who is actually a patent attorney, by education anyway. He didn't have a real chance to start his career when he finished his degree, before the marbles got him and he was shanghaied into the Army. Those dreaded marbles with the numbers, dropped into a barrel once a month. If your birthday falls on that day, you're called up—that's that. Tough luck! If you're a student, you can defer. But they'll get you anyway, later! Next to Geoff is another Scheyville bloke. There's three of us brand new Second Lieutenants. We are the only officers in this bunch of 40 or so men.

There's a smell in the air. Not sure what, tropical? It's the moist air carrying the camphor and canvas smells emanating from the tents all around us? I realise suddenly that there are rows and rows of tents. There must be a shit load of courses going through JTC at the moment. Wow! Funny that I didn't notice any of this last night. I barely remember how I got here. We were piled onto buses at the airport and came here after dark.

I'm dreaming, but then:

'Right turn! By the right... quiiiiiiick... MARCH!... doubllllllle... MARCH! On the double, you slack arses! Lef-ri-lef-ri-lef-ri... get into step, keep up at the back!'

Jeez, these guys mean it. I'm barely awake. Perhaps we're running to breakfast... no, bugger... we're running straight past the mess hall. We're running across a river on a swing bridge, whoa this thing is swinging!

'Break step slackos... do you want the bridge to start swaying and tip you all into the river?'

Okay, okay, so how are we supposed to know? We sort of stumble out of step across the bridge, still in a hurry, then...

'Get into step people, who do you think you are? Lef-ri-lef-ri-lef-ri!'

After a few minutes, I understand why this place is called the Jungle Training Centre. The rainforest is closing in. I think we are slowly going uphill, yes, I can feel it.

'Welcome to Heartbreak Hill you lot!' one of the instructors yells.

Actually it's not too hard. We're just running in our greens, no webbing, no packs. I am used to running at OCS with all the gear on, so this is okay.

'If you think this is easy fellas, in a few days we will be doing this every day, with your webbing plus rifles and machine guns... just thought we'd let you'se know. Gives ya something to look forward to!'

Bugger, why did I have to think it was easy. The NCOs must have read my mind! I can hear Geoff mumbling something under his breath next to me. Wonder what he's thinking?

Finally, we make it to breakfast. Thank God, I'm starving! The food is bloody good here! Brekky is over in no time and I barely have time to think. Its go, go, go... all day, until we collapse in our tents at night.

Time passes fast. We've been here a week now. Tomorrow they reckon we'll do the obstacle course, a nasty place... with pipes to crawl through and ditches to jump and walls to climb. Looks really bad. At the end you have to fall into a huge mud-filled hole they call the bearpit: filthy stinking mud! I've seen blokes from the other courses do it and pop up looking like wide-eyed maniacs, dripping and oozing this shitty looking mud! The whole thing finishes when you climb the tower and jump into the Canungra River. Hot sweaty bodies into a cold river!

But now we're running up Heartbreak Hill carrying webbing and rifles, with the three leading soldiers in the squad carrying an M60 machine gun (MG) each. They're heavy! Every 50 metres or so, the instructor yells out 'change' and the MGs are taken over by the next three. That's okay. But now us three officers are running abreast and now we're carrying an MG each. The red headed sergeant is saying nothing. We've done our 50 metres but still he says nothing. Hey, come on, you're supposed to yell out 'change' after 50 metres or so, remember? I'm thinking he'll make us carry the MGs for the rest of the way up the hill to the top. Jeez, that's still hundreds of metres away. He's not going to beat us. We're still running. We're just about on our knees! But, we make it! Then, loud enough for everyone to hear, the Sergeant says:

'Well done, Sirs. Pass the MGs on now'.

He just thought he would see if we would break and was big enough to acknowledge our effort. We feel good!

Later, down at the live-firing attack range, we are waiting our turn. What an amazing scene! They fire live 7.62mm rounds over the heads of the soldiers, from a Bren Gun. It's mounted to fire into a large cliff face into which the rounds ricochet. Pyrotechnics are detonated in 'safe' areas to simulate mortar attack, and smoke bombs go off all around the place. The effect is a cacophony of ear splitting explosions, whistling bullets and shattering rock, mixed with the pungent smells of gunpowder. All this to simulate real battle! Now it's our turn and I've been nominated as the Platoon Commander for the attack.

'Okay now, do you see those earthen bunds everywhere? Make sure you break up the attack so that there is always covering fire from at least one of your sections as the others are running forward, and never repeat the same pattern!' says the instructor. He's a big bloke. A Lieutenant.

Right, so here we go! My first action with real soldiers. I bark out orders for covering fire and we charge from bund to bund until we get to the other end. Even though the firing is not aimed at us, it bloody well feels like it is, as the bullets whistle just over our heads!

'Why didn't you listen to me, mister?' an angry voice shouts.

In the noise and mayhem, I'd been feeling cocky, but suddenly I was brought down to earth. Surely, he's not talking to me? He is!

'I said do not repeat the same pattern in the attack when ordering your sections forward. But that's exactly what you did! By being predictable, the enemy would have known where to direct their fire! If this hadn't been a simulation, soldiers could have been killed. Remember that, mister!'

Another lesson.

It's now our last week and we're at a place called Levers Plateau. Its infamous for being cold and wet, full of leeches and ticks. For four days we've been miserable, wet to the bone, hungry and tired. They barely let us sleep at night and keep us alert with endless simulated enemy probes. What a finale!

Finally, our three weeks at Canungra are over. It has been a life-changing experience!

We headed off on our posting to 4th Field Regiment, but not before the red-headed Sergeant had come up to the three of us and shook our hands to wish us good luck. He looked us in the eyes and reminded us of our responsibilities as young officers.

'The young diggers look up to you Sirs, remember that. If you let them down, they might die, or some of their mates might die. They don't forgive mistakes, you know what I mean?'

His voice trailed off. A sobering thought. We knew what he meant.

4th Field was due to replace 1st Field Regiment, battery by battery, starting around March 1970. The regiment had been training for their specific Vietnam role for over eighteen months and the four of us who joined from the 2/69 YOs course were the most inexperienced of all the officers in the regiment. However, for reasons that I still don't understand, three of the four of us were to be the very first from the regiment to be deployed. We 'marched-in' to 4th Field in Townsville in December 1969. Christmas was just around the corner and what we really wanted was to be with our families. It had been a physically and mentally challenging eighteen months and we just wanted some time out. But the Commanding Officer (CO) had different ideas and we were told to cart our gear to the mess.

The Officer's Mess in Townsville stood bold and shadeless in the new Lavarack Barracks, where no tree was any higher than a metre. The curlews cried out by night and the air shimmered in the heat by day. Behind the barracks stood Mount Stuart, a forbidding rocky face visible from every part of Townsville. My bed, like all the other beds, had a plastic covered mattress. Apparently, some quartermaster thought it was a good idea for the Army to buy plastic covered mattresses, so the soldiers wouldn't soil the fabric with their sweat. There was no chance of that. We floated on our sweat instead, lying there during sleepless nights, listening to the screeching baby-throttling cries of the curlews and the swooshing sound of the overhead fans. It was the CO, a Lieutenant Colonel Forward, who delivered the stunning news. He was a swarthy looking man in huge Bombay bloomer shorts and a handlebar moustache. He reminded me of the military types that you would see in old movies of the Indian Army, kind of 'pucka' in a pompous way. Three of the four of us newcomers were summoned to his office.

'I am going to send you boys first. You will go and work with 1ˢᵗ Field Regiment which is in Vietnam now, and be my 'vanguard', and make sure you are acclimatised and ready for us when the regiment is deployed in March. You will leave in the first week of January, and between now and then you will be on exercise in Shoalwater Bay to get yourselves ready,' he said. I couldn't believe my ears. What's a vanguard? What are we supposed to do there? Should he not be sending more experienced officers first? He has to be having us on, I thought. But no, apparently not. What was his reasoning? Even years later I was not sure. That group of very tightly knit officers that I was to be part of in 4ᵗʰ Field would be the group that I would continue to serve with over the next few years. Strangely though, I would never be considered a part of them. But more on that later.

I remember little of that exercise at Shoalwater Bay. It was a military training area that I would have much to do with in later years, but for now all I wanted was to go home to Brisbane and spend Christmas with my family, and then head to Sydney to see Lo. All that passed very quickly, so here I was on the 7ᵗʰ of January 1970 in Vietnam, on an FSB in the hands of one of the gun sergeants, about to be introduced to the life of a gunner officer in Vietnam.

'364 sleeps and a greasy egg' was the standard greeting for a fresh face like me. The term derived from the fact that a normal tour of duty in Vietnam was one year: 365 days, usually to the day! I quickly learned what that meant as everyone I met would always point out just how many sleeps were left before their RTA (return to Australia). Naturally a new 'reo' always had more days to go than everyone else who was already there, and some of them would even cheerily announce that they were on 'happy pills'—Quinine tablets that all soldiers were given 14 days before RTA to knock out any malaria that might be in the system. The daily tablets of paludrine that we took only suppressed the malaria, but did not remove it from our bodies.

Being in a theatre of war and having to live and breathe the daily grind of operations, whether planning an action, executing it or recovering from it, is entirely different to learning about it in a classroom or on a course, such as the one at JTC. At first, the routine did not seem too hard. It was the job of the gunners to provide fire support to the infantry patrols, which were out and about amongst the rice paddies, villages, rubber plantations but mostly deep in the jungle, in what was referred to as the 'Area of Operations' (AO). The infantry would find and destroy or capture the enemy and generally make it difficult to move, re-supply and execute their planned modus operandi. The enemy was the North Vietnamese Army (NVA) supported by the local militia, known as the Viet Cong, or VC for short. The NVA were the regular soldiers of the revolution. These were mostly highly trained and skilled fighters, well organised into regiments that reported to a chain of command that issued orders and instructions all the way from Hanoi. They were deployed south to carry the fight to the Americans. They cared little whether the enemy was American, Australian or some other nationality. They knew that whoever was in their country was an invader, an aggressor and had to be removed. The NVA and the VC moved quietly. They rarely if ever used roads. Their home was the jungle, in which they skilfully moved about, avoiding tracks... silently, mostly at night. They were a determined enemy: insurrection was their aim. An insidious and difficult enemy!

VC lived everywhere, some in the villages amongst their own people; some more hardened ones lived in caves, or in mountain camps, in heavily defended and dug-in complexes that included schools and hospitals. As the villagers of the south became sympathetic to the cause of communism, they were coerced into providing their sons and daughters—and their fathers— and food to the VC.

The Aussie way of bringing the fight to the enemy was to take to the jungle in the same way that the VC did. This was unlike the US Army, which rarely went anywhere without preliminary bombardment, helicopter gunship support and general fanfare and fuss, only to sit in FSBs and not go 'beyond the wire' unless in large numbers and with much noisy fire support. The Australians adopted a well proven method of guerrilla warfare, long-practiced in the Malay and Indo campaigns: engage the enemy on his home turf where he thinks he is untouchable, isolate him from his local support and create friends in the local community by winning the battle for their hearts and minds.

For the Australians, this meant patrolling for weeks and months at a time, never walking on tracks that may be mined but paralleling them, in an exhausting effort to avoid being ambushed by the enemy, yet find and engage him. The Australian patrols would stay out in their AOs weeks on end and be re-supplied at irregular intervals by Iroquois helicopters, sometimes with significant gaps of up to 10 days. This meant carrying large quantities of ammunition, food and essentials such as water purification tablets. They would stay out so long that their battle greens would begin to rot off their bodies from the constant sweat, and frequently the re-supply helicopters dumped sandbags full of fresh clothes, new boots and socks. The men would strip and change on the spot and the putrid, discarded greens would be returned to Nui Dat to be burnt.

Each infantry battalion had a number of artillery Forward Observers (FO) with them. It was their job to 'call' supporting fire from the gun battery, or the whole artillery regiment if necessary, and any other fire support assets that may be in range or available. The FOs would coordinate all fire support, including not only their own artillery units, but also mortars, air to ground, and Naval gun fire. Air to ground included support from the US F4 Phantoms and helicopter gunships. 'Artillery' included all artillery from the Australians or the Americans that may be in range, from 105mm howitzers to the 155mm and 175mm guns, the big 8-inch guns, and any of the mortar companies. Any destroyer off the coast could provide Naval gun fire. Being an FO during a 'contact' with the enemy was potentially a very complicated and exacting task. It was vital to know exactly where all the friendly troops were and just where the enemy was, so that the supporting fire could be brought to bear accurately on the enemy but safely away from friendly troops. It was also a task that demanded absolute and exact timing, with every second vital, including the time that the rounds would travel in the air from the moment they were fired at the gun line to the moment they detonated on the target. This was critical to troops that may be forming up to stage an assault. The timings, directions and trajectories of all the fire support from every gun position would also be transmitted to all aircraft in the area, enabling pilots to take avoiding action.

Later in my career, when I became a senior FO and then a Battery Commander, I was to learn in great detail the art of what is called 'deliberate fire planning'—how to coordinate fire support from all these sources. It became my bread and butter. However, at this early stage of my career, I was too inexperienced to be deployed as an FO other than on short patrols. My role was to ensure the gun line operated properly, the gun detachments were doing their jobs, ammunition was prepared and the guns maintained. During fire missions I ensured the guns were aimed, or as we called it 'laid' on their target with the correct bearing, elevation and ammunition. I would also take my turn in the command post during the quiet hours to act as a relief Gun Position Officer, or 'GPO'. The GPO's role was to manage the command post and supervise the operators doing the computations required to ensure the guns engaged the correct targets, then order the guns to engage. No computers in those days; just graphs, slide rules and tables.

As a young officer there were lots of other jobs such as organising work parties, planning deployments and looking after the welfare and health of the soldiers in my section. The Vietnam war is well known for its use of Natios whose numbers made up about half of the regiment's strength. The majority of these men were happy to be there, but there were some who just wanted to go home. The Natios that I dealt with were mostly keen young men, happy to do their military service. They were often university graduates whose service was deferred until after graduation. Thus, the educational standard of these Natios was quite high. At one stage I worked in a command post with a science graduate on one side, an engineering graduate on the other, and a highly trained radio technician behind me: very intimidating to a young officer who had no university degree. Later I learned it was not unusual for an officer to be supervising soldiers or civilians who are more qualified than the officer in specific areas. An Army officer is a leader and a manager, trained to bring out the best from the wide variety of skills and qualifications under his or her command.

Frequently the daily routine on operations could be boring. We played volleyball or kicked a footy from one end of the FSB to another. One game of footy was suddenly interrupted by a misfire from a tank that was parked nearby. It was memorable! There we were, doing some punt kicks about the place, when suddenly there was one almighty bang and the sound of a projectile whizzing past our ears. We were shocked. No-one knew what happened, but we all stared in disbelief at the centurion tank that was parked in a defensive position in the centre of the FSB. After a few moments, the turret hatch popped up and a head appeared.

'Sorry fellas, misfire!' We stood stunned.

Surely a tank doesn't have a misfire, especially when the barrel is pointed towards its own men who happen to be off duty kicking a ball around? But hey, this one did! Apparently, the tank commander had made a mistake when cleaning his gun. In Vietnam, tanks, like all weapons and weapon systems, were loaded ready for action most of the time, unless they were in safe, rear areas. The misfire came as a huge shock to the tank crew, but an even greater one to us. It was lucky that no-one was hurt. I'm sure that a few heads would have rolled in the armoured corps that day.

They say war is mostly sitting around waiting for something to happen, and in Vietnam, for most of us, that proved to be right. Sometimes the daily routine was nothing more than preparing for the night routine. During the day the infantry patrols would seek out the VC on search and destroy missions or simply hold ground somewhere. At night they would settle into defensive locations, position their early warning pickets and set out ambushes. The FOs would study the layout and most likely enemy approaches and prepare a list of possible targets that they might need to engage during the night to support the springing of an ambush, or an attack by the VC, or simply to light up the sky with artillery flares that hung like huge candles in the sky, turning the night into day, and revealing any enemy movement over open terrain. The most important of the defensive fire targets would be confirmed by adjusting live fire from one of the guns in the battery, so that the information to engage that target should the need arise, would be pre-recorded and ready to go. Frequently the FOs were in dense jungle, unable to observe the fall of shot to satisfy themselves that the target was correctly adjusted. In that case they would do 'sound adjustment'. This meant picking a safe grid reference (location) sufficiently far from their position, and bringing the fire in, one round at a time, using only the sound of the impacting round as a guide, adjusting each gun progressively closer until the fragments could be heard whistling through the leaves above. When the FO was satisfied that the fall of shot was as close as possible without actually bringing it in on top of his own head, he would tell the gunline to record that target, and save the computations. If a particular FO and his patrol were considered to be in a situation of imminent threat, the battery's guns would be laid (aimed and loaded) at that target, ready to fire in an emergency. A term we used to describe such a target was 'final protective fire'.

When an FO became 'lost' (temporarily displaced was the term we used) which was not hard in that heavy terrain, with no modern navigation equipment like GPS, the guns fired White Phosphorous—sometimes referred to as 'Willie Peet'—300 metres above the ground in a triangular pattern, so that the FO could take a bearing on what he saw, or heard, and work out his location. If a US 'callsign' (the name given to the unit or patrol) was in the area, they might call for 'Wilson Picket... at 300'. The Vietnam war developed its own lingo.

On most nights, the battery would also conduct what was called 'H and Is' or Harassing and Interdictory Fire. H and Is were fired at irregular intervals throughout the night at targets selected by the FOs, to include such things as road junctions, potential enemy meeting places, tracks and other locations where the VC might be moving at night. The guns that were used to fire the H and Is would usually be different to the guns that were laid on the final protective fire target. As a young officer, managing the H and Is often became my task those first few months and so I would get up four or five times in the night, wake those gun detachments in the battery that were nominated for the night, and conduct the fire missions. Amazingly, in a demonstration of how quickly your body can adapt, those who were not on H and I duty would simply sleep straight through the night. This, in spite of the fact that at least two or three 105 mm howitzers would be blasting away for twenty minutes at a time, just metres from sleeping heads.

Regularly, US Navy ships sailing offshore would add their firepower to the H and Is. Naval guns are large calibre, six or eight-inch monsters. They pack a lethal load. One day, Major Keith Hall, my Battery Commander (BC) decided that it was time for the naval fellows to learn what it was like on land and vice versa. So, off I went with a handful of others, picked up by a US Navy helicopter, to join a US destroyer whilst a load of US Navy gunners came ashore to sleep in our grubby hutchies. I know who got the better deal. We went from sleeping rough to sleeping in a cabin and eating in a wardroom. They would have been quite shocked spending the night in our weapon pits with sandbags and dirt-covered corrugated iron overhead.

Landing a helicopter on a ship that is powering along in big seas has its moments, especially when you realise that the way they avoid a helicopter smashing itself to pieces against the deck is to connect it to the deck and drag it down slowly, while the helicopter is actually trying to take off. It's a fascinating, clever concept. As the ship rolls and the deck heaves up and down, the ship's loadmaster stands in the middle of the deck's landing zone, right underneath where the heli is lowering itself down. There is not a lot of room for error as he connects a cable from the deck to a hook under the helicopter. Once the lifeline between the ship and the heli is established, the pilot gets a 'thumbs-up' from the loadmaster. This is the signal for the pilot to gun the helicopter's engines and lift the heli gently up and away from the deck. When the full extent of the cable is reached the pilot keeps the strain at a controlled level whilst the cable is mechanically wound down, very slowly, towards the deck. The drill ensures a slow, relatively gentle landing, with the helicopter actually heaving up and down in symmetry with the deck of the ship, whilst being pulled towards it at a controlled pace. It is sheer engineering genius and a marvel to witness. A bit worrisome to be part of though!

The most fascinating event was to watch the ship delivering its H and I support. The Captain invited us to the bridge to watch. We saw the gyro-controlled barrels align flat and level at targets that were deep inland, well over the coastal ranges, and stay aligned in perfect unison even as the ship rolled and steamed ahead, smashing its way through the large waves. I was amazed. I will never forget the shudders that went right through the hull of that massive steel behemoth, every time the 18 guns simultaneously despatched their volleys of 100 pound-plus shells. Incredibly, this went on right through the night, with the rattles and the bangs inside the small cabin that I shared with five others, waking me every 30 to 40 minutes. So, this is what the big ships do when we call in the naval gunfire? We flew back to our grubby dusty fire bases the next day, with greater respect for the Navy!

The fire missions that were really exciting though, were the 'contact' missions. These were fired in support of a patrol engaging the enemy. There was nothing as stirring as hearing the call over the radio: 'CONTACT... CONTACT... FIRE MISSION BATTERY' to get the adrenalin going. The gun position would erupt, everyone running to their respective positions: command post operators diving headlong into the command post, gunners running to their guns, and the section commanders, like me, running to the gun line with the sound of the fire orders ringing in our ears! 'Take post! Take post!' would be the call, and in less than 60 seconds the first rounds would be on their way. Sometimes it would fizz to nothing, but sometimes the fire mission would go for many hours, even days, with perhaps sporadic breaks and then long periods of almost non-stop firing, as the FO sought to bring artillery fire down on the enemy and cut off their possible escape routes.

Preparing 105mm rounds during a break in the action at Fire Support Base Coral.
Photo Commonwealth of Australia DVA

When these contact missions were on, no one was spared from carting ammunition. Everybody, including officers, drivers, cooks and 'bottlewashers' as the term goes, would be up all hours of the night breaking open boxes of ammunition and carting the shells and the cartridge cases to the gun line. For the really big contacts it was often necessary to call for ammunition re-supply and the Chinooks would fly day and night, with underslung packs of ammunition that they would drop as close as possible to the gun line.

Once the contact missions were over there would be a huge mess of boxes and bits and pieces to clean up. Each cartridge case contained seven charge bags full of propellant. However, unless the guns were firing at maximum range, not all of these charge bags would be used. So, there were usually piles of unused bags of propellant lying about that would need to be burned. The bags contained small pellets of cordite, which had another useful purpose; boiling our billies, very quickly! They would bring a cup of water to the boil in just a few seconds. This was done by lighting a small block of hexamine, which is a camping fuel issued to soldiers with their rations, and carefully rolling the pellets under our pre-positioned metal brew mugs, or 'cups canteen' as they were known. The pellets would flash intensely and heat the water almost instantly. It was a great way to make a quick brew, but incredibly risky. If the flash from the igniting pellets was to reach any of the pellets that were held ready in the hand, the lot would likely go up in a big hot flash. Over the years a few gunners were badly burned. Stupid thing to do, really.

Much of my time in Vietnam was spent on the gun line, inside the safe perimeter of a Fire Support Base, but I did go out as an FO a few times, on some of the shorter patrols, when the more experienced observers were on other tasks. I must admit, the thought of stepping on a mine was always there. The VC, by carefully lifting and disarming each mine, had removed all the 'jumping jack' anti-personnel mines from an Australian minefield that was laid in the vicinity of 'the Horseshoe' a few years before I arrived in Vietnam. The purpose of that minefield was to protect one of the approaches to Nui Dat. Tragically, those mines were used against the Aussies—very effectively. The minefield was a catastrophic error of judgement by the Australian commander at that time. It was an overly ambitious idea which ignored lessons from the past that should have been remembered. A minefield must be overseen or patrolled regularly. If a minefield is left unobserved, it permits those whose progress you are trying stop or channel to locate the mines and remove them or render them useless.

Not long after I returned to Nui Dat from FSB Peggy a fellow gunner officer, by the name of Bernie Garland, was killed by one of the mines lifted from this very minefield. Bernie was an FO who had recently been home on R and R to marry his sweetheart and he was on his very last day in the 'long green' (as the jungle was called), at the end of the last patrol of his tour of duty as an FO. He was returning from an operation near Xuyen Moc, a small town on the eastern fringe of the province. When walking less than 50m to the helicopter that was to carry him back to Nui Dat to commence his 14 days on 'happy pills' before flying home, he stepped on a mine. He died right there in the arms of his Bombardier. How ironic that after all the steps he took in that country, with only a few steps to go, he stepped on a mine—one of the many mines originally laid by Australians. The folly of war!

By May of 1970, all the batteries and personnel of 1st Field Regiment had returned home and been replaced by my regiment, the 4th Field Regiment. It was my turn to gloat as all the new comers had 365 to go, whereas I had only about 250 to go! Now, I was the experienced young officer, with the lessons of the past few months ground into my head, many of them forming the basis of habits that were to be with me for life. Good habits hopefully, such as those that help to keep me safe in the bush. The NCOs of 1 Field, tough and callous, jaded from the daily grind of war and hard on me as a newcomer, had taken me under their wing. They are a bunch I won't forget.

7 A YEAR IN PHUOC TUY

After the first month, I was no longer a pale body from Australia. I was now brown as a berry from running about the gun line every day in the strong, tropical sunshine and incredible humidity. On and around the guns, we mostly wore just the minimum... 'jungle greens' comprising trousers, boots general-purpose, bush hat and often no shirt, but always with a weapon by the side, in hand or just nearby. The loaded weapon, referred to in the Army as 'your personal fire-arm', whether it be a 9mm Browning pistol, a 7.62mm Aussie SLR, or a 5.56mm US Armalite, was like an extension of the body. It felt strange whenever we were not holding a weapon. Always loaded with safety catch on 'safe', more than a few soldiers accidentally shot themselves in the foot, or worse in the head, through carelessness. By now I voiced all the right 'in-country' terms and had taken up that wiry, slender look that most servicemen in Vietnam had, caused by the hard work and 'hard' rations that we were on. Ration packs only go so far in satisfying a body's daily intake of required calories and vitamins. I still had not seen anything much outside of 'the wire' at FSB Peggy, apart from a few short operations on overnight ambush patrol as an FO. That was quite a nervous time, being asked to act as an FO the first time. Having only ever controlled a fire mission in the safety of a gunnery range in Australia, I suddenly became aware of the need to be spot-on with everything as the lives of the platoon may depend on my

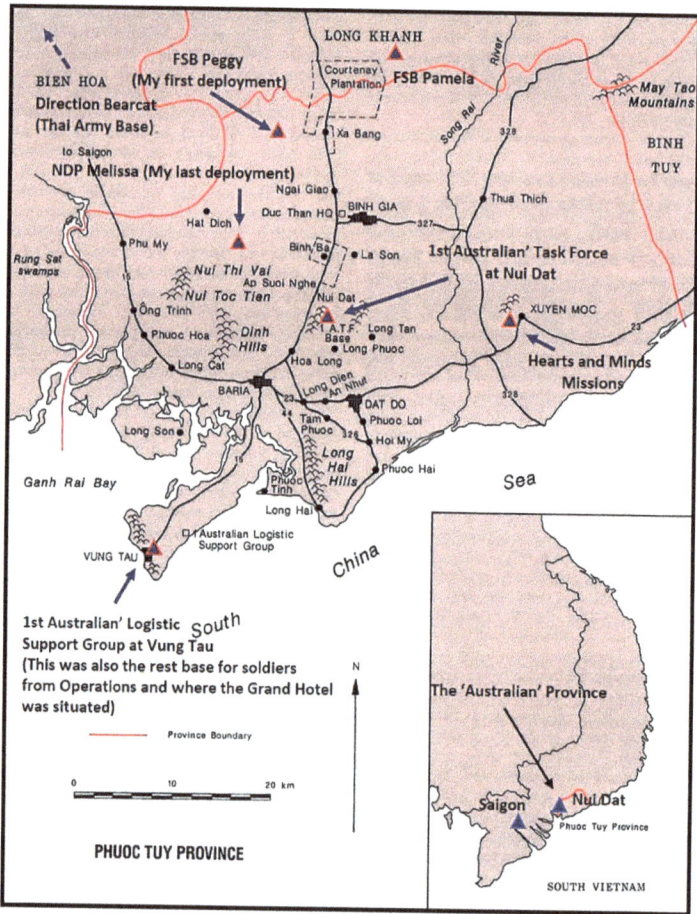

The First Australian Task Force was responsible for Phuoc Tuy Province. Vung Tau was the port of entry and the location of the Australian Logistic Group, which supported the forward elements of the Australian forces, the combat elements, based at Nui Dat.
Map Commonwealth of Australia DVA with author's notations

proficiency as an FO. Thankfully, no VC came wandering along the track that first night. The ambush was not sprung. I was relieved.

I was sitting on some sand bags one day, next to one of my command post operators, generally chatting about stuff, when he put his Armalite down between his knees, butt on the ground, barrel facing upwards towards his face. He smacked the ground with the butt and unbeknown to him, his safety catch was on 'fire', instead of 'safe'. The weapon fired! He shot his slouch hat clear of his head as the projectile passed within a millimetre of his nose, straight through the rim of the hat. We sat in shocked silence. That was close! He was automatically charged with 'occasioning an unauthorised discharge of a weapon' and, as was the way in those days, the CO docked his pay for a couple of weeks. That would not happen these days, because in the case of married soldiers, it would be the family that would suffer.

Sweat seemed to be the order of the day. Everybody glistened, and any clothing or bedding had that permanent sour 'sweaty-body' stink about it. We needed plenty of water, so each of the gun detachments hung several large canvas bags with fresh drinking water that was cooled by the old fashioned 'bush' evaporation system. It tasted a bit 'canvasy', but it was good. I remember the pleasure of drinking that cool water in the dusty hot surrounds of a fire support base to this day! Back in the days before car fridges and freezers, you would see these canvas bags hanging off the bull bars of farmers' Holden utes. They are always cool because they work on the proven principle that the warmer liquid is evaporated first, leaving the cooler liquid behind. It all assumes that you have sufficient water to re-fill the bags from time to time. On the gun line, water was not a problem. We didn't see the sun as a problem either. Nobody wore sun screen!

Back in Nui Dat things were a little more formal. We would wear shirts with badges of rank on them, and suddenly people would salute again. After those first few months 'in country', I was given my first two-day leave. I was still quite unfamiliar with Nui Dat, but even more unfamiliar with Vung Tau, and very excited to have a break from the daily grind and smell of operations. There is a smell associated with the gun line that is etched into my brain: the smell of burning cordite when rounds are fired. Getting away from it to experience the smells of rural Vietnam was, if nothing else, a change, but I soon realised that most Vietnamese villages had a sweet smell of raw sewage mixed with rotting food and open drains. I settled into the Colonial Hotel at Vung Tau. Aptly named and remarkably like a beautiful old colonial hotel anywhere in Asia, the hotel provided a non-military environment for Aussie soldiers who needed a break. I sat and made a tape to send to my parents, and one to Lo. The tapes that we used were small reel to reel tapes that had about 45 minutes of talking on them. Lo and I lived for those tapes! I used to sneak away to the most unlikely locations to record mine, locations such as spare weapon pits or unmanned machine gun posts, whilst looking out over the many rolls of wire that kept us safe. Taping completed, I had a few 'Tiger' beers and set off to explore the downtown bars. Vung Tau was mostly a ramshackle place, with alleys and dark places and thousands of scooters. We were under strict orders as to where we could go and where we could not, and which bars to avoid. The MPs were everywhere!

What a night it was! I had been warned off the pretty little Vietnamese bar girls who would encourage you to buy them a drink whilst you drank your own. However, they were part of the bar scene and the barman would simply pour them a brown coloured cordial but charge you for a whiskey. I ran into some Americans and between us we managed to stumble from bar to bar, mostly avoiding the bar girls. But... clearly, not quite!

I can remember being panicked by the Americans:

'Hey man the MPs are outside... it's after curfew...RUN!'

Out the back we ran, and it was at this time one of the bar girls saw an opportunity to take an 'Uc da loi' home. That was their name for an Australian soldier. Every Vietnamese girl wanted to be taken to Australia or America by a soldier. To get out of this war-torn country was their aim. Some succeeded, but not many. Some even had children by soldiers in the hope of being taken home, but even then, not many succeeded because Australia and the US had active policies that discouraged such relationships.

This girl took me by the hand and led me through back alleys and open sewers and slippery drains and dead cats to a house somewhere in the back streets. House was a generous description for the flimsy woven grass-walled attap structure.

'You miss curfew... you go to jail... you safe here!' she said, in broken American-accented English.

I can remember she was a pretty thing, but then most of the Vietnamese girls were pretty when they were young. The hard life and the climate, and I suppose the genes, would turn most of them into wrinkled 'mama-sans' at a relatively early age—around their forties. She was right about going to gaol for missing curfew, as that was standard practice. Any soldier found out and about after curfew, which, from memory was about 10 pm, was always taken to the MP lock-up, from where he would have to explain to his CO the next morning why he had been so stupid as to miss curfew. I was not so sure about being safe in that house though. The VC were known to co-inhabit many of the houses in the villages, slipping out to fight at night and working in the shops or the paddy fields by day. But I wasn't in any state to argue. What had happened to my two American friends? Had they been similarly rescued into small houses somewhere? I looked around. There was papa-san. Toothless grin but he looked harmless enough. And there was mama-san. Another toothless grin. Oh, and baby-san! Whose baby was that? There didn't appear to be an eligible husband in the house.

She led me to a bed, also made of woven grass, and indicated I should lie down! I did. In moments I was out to it! When I woke and remembered where I was, my first instinct was my wallet. Was it there? Yes, no problem, still in my pocket and still full of Military Payment Certificates (MPC). MPCs passed as legal currency. The US had decided to withdraw the greenback from Vietnam as it became the subject of fraud and speculation. The MPC was the same value as the greenback but had no value outside of Vietnam. I looked about and sure enough, the whole family was there still, looking at me expectantly. Papa-san and mama-san were still grinning and nodding. So, after a cup of ultra-sweet tea, I peeled of a suitable wad of MPC, handed it to papa-san, who gave me yet another toothless grin, and bade my farewell. A night to remember. Not all the Vietnamese were out to get you.

To the west of Nui Dat lay the Thai Army base of Bearcat. The Thais had some 8000 troops in Vietnam. They were supposed to be there to help the US effort. The truth was, they were simply there to please their political masters so that the Americans could boast that another Asian country was helping to stem the flow of communism. The CO had sent me to Bearcat for a period of six weeks to act as the Australian Liaison Officer. I lived amongst the Thais, but in an enclave of American military liaison people, four of whom were assigned to assist me as radio signallers.

In the time I was there, no Thai patrol ventured outside of the wire... not one. However, they made life very comfortable for themselves, with Thai officers building large homes inside the wire, using materials that were destined for the war effort; and they had servants just like back in Thailand. They ate like kings whilst their soldiers were fed on a bowl of rice and a bit of fish sauce. I was befriended by one of the Thai officers who invited me for dinner at his 'house' several times. I went once—that was enough. I found it disturbing to eat a fancy three-course meal, very delicious mind you, to drink cognac and be served by a steward in a house built within a base that is supposed to be a place to fight from, not a place of luxury. I only saw Thai officers in their officers mess or in their fancy houses. After a while I stopped even eating in the officers mess. The officers ate very well there—the Thai food was superb. But in the soldier's mess, the men ate rice with fish paste, full stop! Every day! So, after a week or two I determined to either cook my own food I bought in the Post Exchange (PX)—every American base anywhere in the world has a PX—or eat in the soldier's mess, which I did regularly. Bearcat was like a mid-sized country town and I wondered if there was indeed a war going on.

I did see a lot of 'trade' going on amongst the US soldiers. The place was awash with drugs. After all, the iron triangle, where much of the world's heroin came from in those days, is in Thailand. I suspected that the US units based at Bearcat were not the cream of the US military. In my role as Australian Liaison Officer, it was my responsibility to be aware of all the missions that the Thais were conducting so that if there was any chance of a clash between Australian and Thai forces near the boundary between the two areas of operation, I would be able to warn the Australians and keep an eye on what was going on. I had two Australian signallers under my command plus a small team of US soldiers. I was not at all busy, as the Thais simply did not go anywhere near the Australian boundary. They rarely went beyond the ramparts of their safe surrounds, so to speak.

At night time, I wasn't game to go near the US off-duty soldiers, as they were all usually drunk, or high... or both! The base had bars, restaurants and brothels, hot dog shops and food stalls, and was awash with local Vietnamese. These worked in the kitchens, or as cleaners and laundry maids and it was fascinating to see them enter and leave the base with just a token security check by the guards. I was convinced that the VC had infiltrated this base, which was like a small city, really. My sleeping quarter was an old yurt, with no windows, just mossie screening. It was very insecure. I still wonder how a yurt came to be on a Thai base in Vietnam, but there it was. I was not the only resident of the yurt. Rats lived there with me, along with countless spiders, 'chi chaks' and other crawlies. The rats were pretty social though, sometimes visiting me at night to run across my chest.

My main period of Rest and Recreation came up during this time. R and R, as it was known, was the entitlement of every soldier. The Australian government made use of the US system which provided flights to Asian capitals such as Bangkok, Singapore, Hong Kong and Manilla, where servicemen could stay in military-provided hotels for a five-day break. However, the other destination of the US was Sydney, and Australian soldiers could therefore go home to Australia. That's what I chose to do. I had deliberately waited until August to ensure that when I finished my R and R, I would have the lesser period of time remaining in Vietnam. I wanted to be with Lo, she was all I could think about, so home to Sydney I went!

A few days before leaving I drew a handful of US dollars which was only allowed because I was going on R and R. But no sooner were the notes in my wallet, I was robbed by one of the many Vietnamese women who ventured onto the base. I had foolishly left my wallet unattended for a few moments and I caught her by surprise quickly leaving my yurt. Suspecting that I had just been robbed, I grabbed her as she tried to run past me through the door. As I struggled with her, she managed to secrete a wad of money into some part of her clothing where I was not game to search. So, still hanging on to her arm, whilst she kicked and screamed and tried to scratch me, I yelled for one of the US officers who had run over to see what the problem was, to call the US MPs.

I stood there holding this struggling, screeching woman wondering how long I could do that without me hurting her, or her hurting me, when luckily the MPs arrived. But they also were unwilling to retrieve the money from wherever she had hidden it and then, reminiscent of a scene in a mad-hatter comedy, amidst all the screaming and confusion, and with a growing crowd of increasingly agitated Vietnamese and Thais looking on, the daily clockwork-like monsoon rains broke. Water smashed down! In an instant, we were drenched! One second it was clear, the next second we were standing in a waterfall! This caused complete chaos, as all attempts to conduct an orderly investigation were drowned by rain so loud it obliterated speech. The rain drops were big enough to hurt! The MPs lost their grip, the Vietnamese woman slipped away behind the veil of rain, and disappeared! The show was abruptly over, and I stood there, as did the MPs, and realised my money was gone. I was angry. That base was so slack that they had no way of closing or warning the gates to prevent her from leaving. Then again, nobody had thought to ask her for her ID before the rains came, so we had no idea who she was anyway! The only consolation was that the money may have been used to feed her children or other family members. Hopefully not the VC.

R and R was wonderful. For five days, I was in a place where freedom was just an everyday thing, where people had a choice about their future. Gone from around me were the faces of despair or the eyes of hatred, and, for a short time, I stopped worrying where my side-arm was every time I moved anywhere. Here in Sydney, staying with my brother, I was with my family and with Lo, and I was happy. She was attending Sydney Uni at the time and told me about some of the anti-Vietnam protests but I really didn't want to think much about that. In my opinion those protesters were wrong and simply ignorant of the facts. I thought we were doing an important job over there and they would realise that one day. In the end, years later, it was me who realised who was wrong and what the facts were. There were a lot of lives wasted for political ambitions that were never achieved and dubious to begin with. The R and R days drifted by in a contented haze and then in a flash it was over, and I was once again on the plane back to Saigon.

My term at Bearcat had not yet finished. One of the last things I needed to do was visit a US Ear Specialist. He was not at Bearcat but in Saigon, some 50 km further west. The ear check was necessary after being stunned by a massive ear-shattering 105 mm howitzer round being discharged straight over the top of my head by the gun detachment, as I was stupidly running in front of the gun on my way from one gun to the next to do my pre-fire checks. I was trying to do the gross-error checks on the guns to ensure they had the correct bearing and elevation, when the gun sergeant ordered the gun just behind me to fire. He was well within his rights as the fire was in support of an enemy 'contact' and was urgent. It was my own fault— I was out of position when the guns were given the order to fire. The blast tumbled me over and over and this sent my left ear into a frenzy of whistling and ringing noises. The ringing subsided but returned with a vengeance later in life! The Regimental Aid Post (RAP) had made an appointment with a US Specialist in Saigon but as I was stationed at Bearcat, had told me to arrange my own transport. This was easier said than done, but the Americans came to the rescue and I was taken there in a US troop bus.

The road to the west from Bearcat headed towards Bien Hoa, one of the largest US logistics bases, and then on to Saigon. Bien Hoa was an amazing place. I was stunned to see endless rows of brand new military vehicles, APCs, and helicopters as far as the eye could see. Someone had told me about the incredible US PX that was in Bien Hoa. It was said that you could buy anything there, including a car. They would ship it straight from the manufacturer to your house. I wasn't too sure about that, but I did check the place out for diamonds. Having asked Lo to marry me when I was home on R and R, I reckoned this was the place to get the ring.

'Do we have diamond rings, sir... why of course!' said the little Vietnamese woman behind the counter. 'How big?'

I settled on the biggest one I could afford and paid the little lady in the PX rather gingerly, not sure if I was buying a diamond or a piece of glass. It turned out to be a diamond of high quality and Lo wears it still! I didn't buy a car but did order sets of crockery and cutlery to be sent home and we still use some of them.

Saigon was mind-blowing. Not having experienced an Asian city at that stage of my life, I was stunned by the waves of scooters that advanced upon me from every direction. Despite the war, there is no doubt Saigon was beautiful. I settled for a beer and a steak on the balcony of a French restaurant. Was there a war going on? From there I wandered to the President's Palace and stood in the gateway, the same gateway that I recognised on the news three years later. It became part of a very symbolic scene where a North Vietnamese-driven Russian T54 tank, slowly crushes the gate and inches its way into the President's compound. Watching that scene was the catalyst that suddenly made me realize the futility of that war. As I watched the gates being crushed under that tank, it hit home that we had wasted our efforts and wasted the lives of our mates for a doubtful cause. But for now, I held those iron bars in my hands and looked at the President's Palace, imbued with the importance of our mission.

What was also memorable about that trip was how I made my way back. I had reached Saigon officially and on duty, but how was I to get back? The RAP had made no arrangements. I knew that I should probably report to the Saigon Australian Army Movements Office, which would undoubtedly have put me on the next Hercules to Nui Dat from where I would have to arrange transport to Bearcat. This process could take over a week. So, after talking to an Australian officer who was detached to the US Forward Air Controllers in Bearcat and who was well briefed in the movements of US aircraft around the south of Vietnam, I decided to hitch rides on 'Huey' helicopters.

The Americans flew the ubiquitous Hueys everywhere. I simply walked up to US pilots at the military airfields and asked them where they were going. They were always amazed to see and hear an Australian soldier. The skies above Vietnam were filled with helicopters, so getting a ride was not hard. They were memorable rides, not the least reason being that the Hueys had no seats or seat belts, so I travelled by sitting on a bare metal floor, straining to keep my backside from sliding around the back and straight out of the doors, both of which were wide open. That was the standard 'load ready' fitting for many of the helis. The pilots were mostly young American fliers, quite relaxed and confident young men. However, one set of pilots was particularly slack. The senior pilot was munching a hamburger he had picked up at the mess and the other was drinking a milk shake, as they casually lifted off the ground. I was given a head set, so I could listen to the radio chatter, and all I could hear over the radio was the continual broadcasting of artillery fire missions. I knew about such things because whenever we fired an artillery fire mission, the standard requirement was to broadcast to all aircraft, the locations of the firing point and the target point, and the vertex height of the trajectory. This was to allow any helicopter within range to avoid flying into projectiles. Helicopters were not required to respond. There were so many fire missions and messages being broadcast, almost non-stop, that all the pilots could do was listen to see if any of the fire missions would be likely to affect them and avoid that area. There were countless helicopters in the air at any given moment, all over the skies of Vietnam, including this one I was hitch-hiking in.

These pilots of mine seemed to care little, as one of them casually wrote the coordinates of the fire missions on the inside of the helicopters windscreen, using some sort of thick marker pen that seemed to obliterate his forward view. So many other notations had been made on the windscreen, that it was almost impossible to distinguish which ones were still relevant. It seemed that as soon as the message was scrawled on the windscreen, it was ignored. At one point, on looking out one side of the open doors, I saw a 'white phosphorous' round detonating in mid-air, apparently not more than a hundred metres from the aircraft. Neither pilot seemed to notice. I recognised the kind of mission being fired. It was a marker mission, where the guns would fire three white phosphorous rounds at the points of a triangle to help FOs locate themselves. I anxiously waited to see if the next 'Willie Pete', which is what the Americans had termed WP, would be fired right near or on top of us. But nothing further happened.

The first pilots dropped me off at an unknown location, an air base somewhere to the east of Saigon, not far from the Cambodian border. They told me I would have to ask other pilots for the next leg, which I did, only to find out in mid-air that the next Huey was going via an air base in Cambodia. As we landed the pilot told me to stay in the aircraft as Australians were not supposed to be there. I could well believe it! Had the Australian Army known that I was in Cambodia, even for a short stay, I would have been in serious trouble. But, we flew on safely and after another change-over at another air base, I finally hitched a ride to Bearcat. When I was discharged from the Army, I looked for the doctor's report in my medical documents, to prove that my hearing was affected by my service in Vietnam. Although I had passed the report on to the RAP, it was not on my record. Fortunately, the Department of Veterans Affairs accepted my story without question.

Towards the end of my stay at Bearcat I decided to get me some of the 'u-beaut' US military gear, such as their far superior backpacks, and swap my Australian slouch hat for whatever I could get. Slouch hats were very high in the value stakes of US-Aussie bargaining negotiations and I knew I was in a strong position. However, even with this knowledge, I was astounded when I was offered an M60 machine gun as a trade.

'Hey man, just take it home. No-one will notice brother. Hide it in your gear!' said the US Quarter Master Sergeant.

'Yeah right!' I said.

I resisted and took two backpacks instead. Crazy yanks!

On my last day in Bearcat, I watched a column of US troops that had formed up to leave the base on a patrol, refusing to mount up and move into a formation for deployment outside the wire. Their commanding officer nervously trotted up and down the line trying to persuade his men. The soldiers were unwilling to risk their lives for a war that their people back home no longer believed in. Although I didn't blame them for that, an army cannot condone such behaviour. It was a sickening sight and not representative of the fine US soldiers I came to know in later years.

In contrast, life at the Australian base at Nui Dat was ordered and disciplined—well mostly, anyway. The base consisted of a large hill feature in the centre, which was the actual hill the map referred to as Nui Dat, but as it was the home of the Special Air Service (SAS), we called it SAS hill. The perimeter of the whole base was heavily wired: seven rings of concertina wire, about 20 metres between each ring, with mines in between, and towers with large sweeping searchlights on top kept the bad guys out. The whole perimeter was protected by the interlocking arcs of machine guns mounted on watchtowers. Other than a few interpreters and local officials, there were no Vietnamese civilians inside, and thus no chance for the VC to infiltrate the base. Each Australian soldier carried his loaded personal weapon at all times, in case of an attack. The area was divided into zones, with each unit responsible for the security and management of a zone. The actual base itself was about four by five kilometres in area.

Drinking was done in the usual Officers, Sergeants or Other Ranks messes, and was rationed at two cans per man per day... perhaps! The beer ration was a point of contention, because the number of cans was calculated on the unit personnel strength, so no matter whether the unit was out on operations and not drinking, or indeed whether one person was absent or present, there would be two cans available per day for every person on the nominal roll. Clearly, that ensured there was always plenty of beer.

When not on operations outside the wire, we lived in large, smelly, canvas tents, similar to the ones I had first slept in whilst on my familiarisation course at JTC in Canungra. The tents were sandbagged to waste height all around, in case of mortar attack. They were smelly because in that hot, moist, tropical climate, any material like canvas would quickly become mouldy, as would all our spare clothes. Our sets of dress uniform and civvies were carefully wrapped and sealed in plastic, ready for the day when we would go home. The sandbags were excellent breeding places for mice and rats, and many a night I was aware of little feet running around. I shared my tent with a mongoose. Perhaps it is more accurate to say that that the mongoose shared his tent with me! He, or maybe she, would look at me as if to say:

'Who the hell are you invading my space?'

Mongoose are ferret-like creatures the size of a very large cat, with big teeth and a nasty disposition. The one in my tent would sit on the sandbag wall and stare at me, and then start doing laps around the wall, just as I would try to go to sleep. I reckon he would do at least 20 or thirty laps every time I looked at him. Apart from managing my mongoose, managing mossies was also a major concern. Mossies carried malaria! When on operations, we used mossie propellant so savage it would eat into our greens. I still wonder what it did to our skin! But when we were at Nui Dat we slept under mossie nets, which added to the claustrophobic nature of the smelly, dank tents. I vividly remember the smell of those steamy nights, lying there wet and dripping with perspiration.

At night, when not on duty in the command post, I would spend my spare time writing letters to Lo. That was the highlight of life in Vietnam—receiving and answering mail, but especially from Lo. Sometimes a letter, other times a tape. These were about 45 minutes in length and we always had a one or two in transit. In those days, it took about 10 days for the mail to do its thing, five days each way. I remember getting cranky with Lo one time because she insisted on writing to me about this US serviceman called

Fred Self. He was on R and R in Sydney. I guess he got under my skin as I still recall his name. He must have taken a liking to Lo and apparently chatted freely to her about all his woes and worries of service in Vietnam. Lo was completely oblivious to the fact that I detested hearing all about this other bloke, letter after letter. I decided to spit the dummy and refused to write her any more, naming this fellow 'Self' as my reason. She was caught unawares and wondered why I was being so stupid. But I held out. No letters from me for three weeks. I reckoned I was punishing Lo, but my mates all reckoned I was punishing myself because apparently, I was in a foul mood the whole time! In the end, just as I was about to relent, Lo wrote and wondered in her typical straight forward way, what I was being so silly about, and could we please resume writing?

Thank god I thought... Phew! Back to normal!

It was always good to be back at 'the Dat', as we called it, because of the relative comfort. On operations, we slept wherever we could, in improvised sleeping spaces partially underground in an FSB, or on a wet sleeping mat under a hutchie that was hastily tied, or simply under the stars. Making noise when setting up night defensive positions was out of the question. But when we were at Nui Dat, life was generally about servicing the guns and recovering the equipment from the last operation and getting ready for the next operation. It was also a time to relax. We would usually drink a can or two in the evening, but sometimes we would stray to other messes and drink their beer.

Within the regimental area, there was an area dedicated to the regular US visiting battery, normally an 8-inch or a 175mm gun battery. These batteries were allocated by the US in support of the Aussies. Their guns, much bigger than our 105mm howitzers, were incredibly loud when they discharged their rounds. When they fired as a battery, they would set up an amazing underground blast wave that would travel for hundreds of metres. Our toilets in the regimental area at Nui Dat, were simple complexes of 'thunderboxes' placed over large communal dunny holes. A thunderbox is a metal can with a lid and no bottom—ready to be placed over a pit. For a permanent battery location like the gun emplacements at Nui Dat, these pits, affectionately referred to as 'shit pits', were sizable affairs with many thunderboxes lined up quite close to each other. The phrase 'cheek to cheek' applied to both the thunderboxes and the people occupying them. When one such pit was filled with human excrement, the boxes would be removed, the pit covered over, a new pit dug, and the thunderboxes placed on top again.

It's not hard to imagine that going to the toilet was a very public event, as up to a dozen or more of these thunderboxes would be lined up with just a few inches between them. The whole process of 'going to the crapper', had a certain level of social intercourse associated with it. However, there was also an element of risk, none greater than when the US big guns started firing. This could happen without warning. The underground blast wave would hit the shit pit and all the lids of the unoccupied thunderboxes would leap open in unison, ejecting used toilet paper that would flutter into the air like streamers at a ship's farewell. If you were unlucky enough to be caught out sitting on a thunderbox at the time, you could have your bum tickled by someone else's toilet paper, not to mention the waft of stench that would simultaneously exit all the thunderboxes and to assault your nostrils. Whenever the US boys started firing there would be this mass exit from the thunderbox shed, with blokes trying to hitch up their pants from halfway down their knees! War is hell!

Amazingly, Brian Swift, one of the bombardiers in 107 Battery, the battery that I was assigned to towards the end of my year, returned to visit Nui Dat in the late 90s, on a tour arranged by the RSL. During the visit, he was able to locate the pit that was in use towards the end of his stay. When he left, he had placed his wife's letters in a large plastic bag and flung them into the pit. On return 30 years later, after a short dig down, he came upon the bag and her letters. They were almost unaffected by their submersion in human faeces, which had by now turned to soil, protected by the plastic bag.

This is typical of the tents we used as our sleeping quarters when at Nui Dat. The circuit for
the mongoose to run around was the top layer of sandbags.
Photo Commonwealth of Australia DVA

We ate well when we were inside of our Nui Dat base. The messes received plenty of good quality fresh rations from Australia. When not on operations we became accustomed to eating bacon and eggs for every breakfast and steak for lunch and dinner if we wanted it. However, out in the FSBs, our diet consisted mainly of a variety (interpreting that descriptor liberally) of Australian or American ration packs, supplemented once a week with a batch of hot food cooked in the kitchens of Nui Dat and delivered in Huey helicopters in 'hot boxes'. The ration packs played havoc with a soldier's internals, as they comprised entirely of what we referred to as hard rations. Sometimes these were 10-man rations, which contained larger portions suitable for cooking in a pot for a larger group of people, and sometimes they were 24-hour rations.

In the Australian packs, the hard rations included cans of meat such as the ubiquitous bully beef, or veal sausages, and tiny cans of cheese or vegemite, and breakfast biscuits so hard the soldiers would joke that to eat them you first had to tie them to a rock, immerse the biscuits and rock in water, and eat the biscuits once the rock softens! Sometimes we would get

Aussie 24-hour rations. The cans had to be opened with the FRED... or 'F**king Ridiculous Eating Device'. A number of these are still in use in our kitchen today because they are incredibly well designed and made from the best steel. They remain the only device Lo will use to open cans. Sometimes we would get the US rations, called combat or 'C' rations, which were crafted to suit the American taste: disgustingly sweet with sugar in everything. Americans have weird taste, such as peanut butter and jelly neatly layered in the same can. The Americans also gave us supplementary packs with items like pens and writing paper, toothbrushes and toothpaste, shavers and packs of foul tasting American cigarettes, like filter-less Camel and throat destroying, dry Marlboro. And yes, the infamous American chewin' baccy. None of us were familiar with chewing tobacco: it seemed like a disgusting habit. We all tried it once before spitting out the muck as a blob of yellow, intensely strong, pure tobacco that played havoc with your mouth and your throat. How could anyone possibly like having a lump of tobacco in their mouth, making your teeth yellow and your breath stink?

The US gunners kept to themselves most of the time. However, on one particular evening, the Americans invited some of the soldiers from our unit's headquarters for a drink at their battery boozer. Among them was a young Australian Bombardier who managed the orderly room. He was a clerk: a classic 'pogo'—one who was never expected to cross the wire to experience the dangers beyond the safety of the defended perimeter. He had a young family back in Australia and his young wife would have had no suspicions that he might have come into harm's way. Unfortunately, during the evening, and no doubt after many Budweiser beers, one of the US soldiers gave him a souvenir to take home. It was a hand grenade—a live one. He put it in his pocket. At some time during the night he must have dislodged the safety pin, perhaps in his sleep as he rolled onto the grenade, or perhaps as he woke up to wonder what that lump was that was in his pocket. No one will know exactly, but the grenade detonated. He never knew what happened.

I was given the grisly job of supervising the removal of his body and the clean-up, but luckily the detailed work of collecting the remains was done by a military special investigations unit. I remember vividly his Battery Commander, Major Keith Hall, who was inconsolable and in tears. It's such a tragedy, losing one of your men in this way. It would not have been easy to explain to the bombardier's wife how her husband was killed.

My life in Vietnam continued to be one of small assignments here and there in the regiment, filling in for officers that were assigned elsewhere. Or, I was doing jobs for the CO as some sort of CO's Mr Fixit! This was because I had never been a part of the regimental build-up to the Vietnam deployment and did not fill a permanent position on the regiment's Order of Battle. I was a re-enforcement, and, as a result was frequently placed in temporary command situations. I was fortunate to be posted on active service at all since the war was starting to wind down towards the end of my year in Vietnam. Many in my class missed out altogether. One of the jobs the CO asked me to do was to act as the protective platoon Guard Commander, to manage the protection of a team of Defence and civilian professionals that was dispatched by the brigade to conduct what was referred to as 'hearts and minds missions. I would select a platoon of soldiers from the gunners of the regiment, and we would provide protection for the 'hearts and minds' team.

These teams, comprising Army and civilian doctors, nurses and welfare workers, would go out to the outlying villages where the Australian influence was not yet as apparent, and try to gain the confidence of the villagers. The teams would distribute food, conduct health and dental checks on the locals, administer various vaccinations, distribute medicines and, to keep the masses entertained, put on a movie. My job was to ensure that everybody remained safe. That was easier said than done!

Villages have dark spaces, deep drains and laneways, and a multitude of places for people to hide. I remember that this was a troublesome task that I did not like, as I relate my story:

Driving along the road to Xuyen Moc I feel unsettled because this is a bad area. It's near where Bernie Garland stepped on a mine and was killed a few months ago. This morning, the unit intelligence officer warned me to keep my wits about me. He said the area has many VC sympathisers because we've barely managed to gain the upper hand around here. Bloody hell... isn't this a bit risky? What are these unarmed doctors and nurses doing here? They must be very brave or foolish, or both. And then he told me to watch out for the local militia because they can be friend or foe. Great! So, what does that mean?

We're entering the village now and I study the lay of the streets, the houses, the town square and its approaches... and the faces of the people! I wonder if they always look this unhappy? The kids look okay, though. Man, it stinks! An in your face kind of smell... rotting food, urine, human faeces, animal faeces... what else? I'm not happy. Look at those rooftops! I reckon they must be flat on top with mud railing walls. So that means we can get up there I suppose? Enemy snipers could get up there too. Must ensure we cover the area from the rooftops. I spoke with the blokes this morning and sorted out the machine-gun teams and the rifle sections. All these blokes are gunners, not infantrymen, but they'll be okay.

One of my NCOs comes up to me:

'Hey boss, seen those weird guys with the white gloves? They're dripping with weapons. I don't like the look in their eyes. I don't trust them. Whose side are they on?', he says.

'Our side', I hear myself saying with disbelief.

We stand there looking at these very colourful vigilantes.

'They're the village chief's militia. They're supposed to keep the village safe from the VC!' I say. 'But I reckon there is a fair chance that most if not all of them side with whoever appears... us Aussies or the VC. Can you imagine these show ponies actually taking on the VC?' I say. 'Unlikely! Let's get our blokes in and we will have a briefing and I will give deployment orders'.

I shake hands with the militia leader, grasping his white-gloved hand and looking into his smiling eyes, whilst taking in the bandoliers of ammunition over his shoulder and the hand grenades on his belt. I give him an arc of responsibility for the night, but I note the need to cover the same area with my own men. Then I check with the doctors and nurses and we select the best spot for conducting the health checks and for showing the movie when it gets dark.

I have no idea what movie they will be showing to these Vietnamese folks. Do we have movies in Vietnamese? I guess so. But there are now a lot of young happy faces about. That's good. And there are quite a few old crinkly faces... and lots of women... but where are the young men?

I walk around and do a recce to see if the area is defendable but that's debatable because there are rickety buildings everywhere, mostly shanties, providing lots of cover for attacking troops. The VC don't like these missions where we try to help the population and ingratiate ourselves, so they may probe the area just to make a nuisance of themselves. But they probably won't attack. That would piss the locals off and might even injure or kill some of them.

I take my NCOs and we select gun positions on roof tops, alongside drains and facing up streets, and I give instructions for all-night pickets and the patrolling regime. Then we start the routine. The villagers come with their sick and the line grows longer and longer as the immunisations begin.

Night falls and we enter a night routine which means extra awareness. Telling the men to take turns to sleep, we try to lie down in mud that stinks of cow turds and raw sewerage, but I don't sleep a wink.

I am thankful when the sun's first light begins to appear and my blokes go around and conduct their clearing activities. The doctors and nurses are done and we pack up quickly to go back to Nui Dat. I'm glad it's over without incident! I really respect those doctors and nurses! I wonder how often they do this?

A few months later, I was in a village not far from Xuyen Moc where I returned on another of these 'Mr Fixit' jobs. Alongside my good friend Mick Boyle, a Captain in command of the survey battery, I was tasked by the Brigade Commander to represent him at a ceremony to celebrate the 25th anniversary of the village priest's ordainment. This village comprised entirely ex-North Vietnamese Catholics. These people had fled the north when it became communist, since religious tolerance was not one of Uncle Ho's strong points. We sat with the villagers on long tables filled with delicacies such as jellied cow's intestines, chicken feet soup, grilled monkey's testicles (or so it appeared) and other similar mouth-watering delicacies. Mike and I understood immediately why the Brigade Commander had asked us junior officers to represent him. I looked around at the proud men and women at the tables, who all chatted away in that melodious Vietnamese language, which sounds as if it's being sung. Neither of us could understand a word, and they couldn't understand English. It made for an afternoon filled with lots of smiling and nodding and exaggerated motions to make yourself understood.

In front of me stood a tall glass containing a whiskey coloured liquid and lots of ice. Cordial I thought. But no, one sip and I knew... Johnnie Walker Red! Whiskey it looked like, and whiskey it was! Every time Mike or I took a sip, the bottle of Johnny Walker Red Label, would be produced and the glass immediately topped up again. This was going to be a long afternoon! But worse was yet to come. I simply could not identify anything on that table that I thought I might dare to eat. This was worrying as I didn't want to insult our hosts. However, in my reticence to reach out and respectfully select a minimum of food for my plate, I had opened the opportunity for my hosts to do so, instead. This they did, but not just with the minimum amount of food. To my horror, I found myself staring at a wobbling pile of greasy, jellied intestines. The whole pile smelled as bad as it looked.

Okay, so let's do this, I thought, grabbing a spoon and shoving a generous spoonful into my mouth whilst forcing my lips not to curl and my nose not to screw.

'Aaaah...shiiiiit...this is worse than I thought.'

I looked up at Mike who at that moment was struggling with a clawed chicken foot and then, whilst forcing myself to swallow, I realised that the hosts regarded my food plate in the same manner as my drink glass... and had already replaced the jellied intestines with another spoonful! Back then, in the late 60s, there were no Vietnamese restaurants in Australia, because the migration of refugees had not yet begun. So, I didn't know that Vietnamese food could actually be very nice. It took me many, many years back in Australia before I persuaded myself to once again try Vietnamese food. It was a memorable day.

A few years later I was sitting in my lounge room in Townsville, watching the news, and I momentarily stopped breathing! There on the TV was a news snippet about a village in South Vietnam that had just been burned to the ground and all the inhabitants killed. It was a village of Northern Vietnamese Catholics that had re-settled in the south. Soldiers from the north, acting on the command of the new Unified Vietnamese government, had sought out the village of northern Catholic refugees and killed them all and burned their houses. I stared at the TV which showed bodies piled up in the village and the thin wooden and straw houses burning. Here were the villagers that had shown Mike and I such hospitality. Only a few years ago I had enjoyed their company, and now I stared at images of their bodies. I felt sick. They had fled to the south to escape communist oppression and to practice freedom of religion. They had sought Australia's protection and invited us to share their table.

Australia had failed them. We just went home and we left them, because it all became too difficult, politically and militarily.

8 WINDING DOWN IN SOUTH EAST ASIA

The year was winding down. Now that my R and R was over, the next big target was my RTA... 'Return to Australia'... words that sounded like heaven! Lo was still at Sydney Uni and was writing to me about the anti-war moratoriums, the demonstrations against the war that had become a regular thing. At times, Lo found it hard to declare that she had a boyfriend who was not only in the Army, but serving in Vietnam. The moratoriums were a strong expression, mostly by university students but also by many mothers (quite understandably), of the Australian public's disapproval of not only the war in Vietnam, but more specifically, conscription. The protestors believed that, not only was it a war far away, with little relevance to Australia's way of life (the threat that communism might reach the shores of Australia if Vietnam fell seemed a long stretch to the imagination of most Australians), but to conscript young men and force them to fight overseas was politically and morally unacceptable to the protestors. Australia's military history shows a deep distrust of conscription, with various attempts by Australian governments of the day, as far back as WW1, either failing or being implemented in a restricted manner, where only limited overseas deployment of conscripted troops was permitted.

The war in Vietnam became so unpopular, tens of thousands demonstrated on the streets. Some unfortunately gave vent to their frustrations by taking it out on the soldiers. Even people who worked in trusted institutions such as the PMG (Post Master General), the 'posties', refused to handle our mail at times. This caused great distress amongst the servicemen in Vietnam, which turned into anger. 'Punching a postie' became a popular saying. Sometimes the wharfies refused to load the ships that carried our 'consumables', our daily supplies, back and forth. One of those ships was the *Jeparit*. It was the ship that carried much of our fresh food and pallets of beer. The government solved that problem fairly quickly by commissioning the *Jeparit* to become the HMAS *Jeparit*—loaded, sailed and unloaded by Navy hands. The beer was safe!

Those of us in Vietnam were more confused than angry about all these protests. We were, after all, simply doing what our government ordered us to do. Why were the protests directed at us? We, including all the Natios and the regulars, or 'lifers' as the Natios like to call us, were proud to be Aussie Diggers, doing what the Diggers had done before us. However, on return home, we found that some of the misguided protestors took things to a personal level by verbally, and occasionally physically, abusing returning servicemen. Some of my friends were spat at when they came home and their response was predictable. Fortunately, that did not happen to me, but when in uniform, I did experience from time to time the disapproving looks of everyday Australians. This was saddening and hurtful, because as soldiers we felt we did a good job for our country. Unfortunately, the Australian government had lost control of the public relations battle, underestimating the impact of a war that would be watched nightly on the TV in the lounge rooms back home.

It is commonly said that we lost the war in Vietnam. Yes, the war was lost. It was lost for many reasons, not the least of which is that we, the 'west', tried to use force to impose our western political system and beliefs on an emerging nation, unnaturally divided and grappling with its identity.

But here in Australia, what the government really lost was the will, the political will to explain to the Australian public what was really going on over there. Right or wrong, on the ground, the Australians were actually winning in Phuoc Thuy Province. We cleaned up the province by reducing VC activity to a minimum. Was it our right to do that? Did the Vietnamese actually want us to clean it up? On reflection, I think not. I believe that the majority just wanted to go on with their life as it was. To the peasants this meant minding their stock and growing their crops, and perhaps sending some of their sons and daughters to the cities, to be educated or get a factory job. This is not unlike any other young developing country. I am convinced they cared little about the government to which they paid their taxes or whether that government was a democratically elected one or a communist one.

What was disappointing to us as service men and women is that we, the Australians, together with 'big brother' US, simply stopped fighting, because our governments did not know how to finish a war that they should never have entered. At a government level, for the US and its allies, it was a war they were destined to lose, no matter how many battles they won. But to us returning Australian soldiers, the war was at a different level. We were the face of the war. We were part of it. We had sweated and fought, and we had killed others. It was a personal thing. Hundreds of young Aussies died over there, and thousands were injured. Most of us had lost mates and saw others become maimed, physically or psychologically, or both. Some of these were Natios, who did not ask to go but did as they were told. We had to reconcile ourselves that it had all been for nothing. That was a bitter thought. But to make it worse, our country let us know that that they were not proud of our efforts, our personal sacrifices. They made us feel that, wearing an Army uniform was wearing a badge of shame.

Many branches of the Returned Services Leagues across the country refused to allow Vietnam Veterans to become members. For years these veterans had to fight to be recognised as having taken part in a real war; one that although politically disastrous and morally unjust, was nevertheless dangerous and life-threatening: for many soul-destroying and for some, life-changing. Each veteran, in their own personal space, was a Digger coming home from a war. But to their dismay they came home to a country, with its proud traditions of military service, that would not recognise them, or pat them on the back and say 'well done son', as it had for all the other Diggers coming home from the wars over generations. Some veterans became bitter men who maintained a rage against society. For me it was disappointing. I was deeply saddened, but I am a pragmatic person and I believed the failures of our government and vagaries of human nature were to blame, rather than anything actually attributed to me personally. I knew we had done our jobs to the best of our capability and that one day our country would recognise that. I believe it has.

My last two months in Vietnam were spent as a gun Section Commander with 107 Battery. Contacts with the NVA and the VC were becoming less frequent. Written orders, found on captured or killed NVA soldiers, told the NVA to avoid the Australian Province on their way to Saigon. The costly battles with the Australians had taken their toll on them, and it was simply not worth their while to engage the Australians. They would leave anyway, once the Americans lost their war, so why waste resources dealing with us? The fact that the tactics employed in our province were achieving successful outcomes was not well known back home. This was something the Australian press did not appear to be interested in. The media wanted the gore of war; sensational stories of Aussies killed or wounded and pictures of dead Vietnamese and body counts!

By this time 107 Battery was split into two independent three-gun sections, and Second Lieutenant Peter Kilpatrick and I teamed up to take one of these sections on the road to support the fragmented infantry operations that were now being conducted. Peter was one of the officers on my course at the School of Artillery, but Peter had not been sent early like the other three of us. So, although I was coming to the end of my tour, he still had about four months to go.

Soldiers of 108 Field Battery positioning one of the guns onto the gun platform, having to push it through the mud.
Photo Commonwealth of Australia DVA

As two young commanders we made a less than auspicious start when, heading up the road on Route 2 to establish a small fire base in the north of the province, we decided it was necessary to overnight somewhere and selected a small, secure compound just north of a village called Binh Bha. The compound was occupied by 'friendly' South Vietnam soldiers. As was the way of the Vietnamese, the soldiers had their families with them. The compound resembled a small village, complete with children who appeared to play with immunity amongst the booby traps that were visibly displayed on the defences surrounding the compound. Barbed wire, supporting auspicious cans of coke, which the villagers told us contained explosives, was randomly strung about the outer perimeter. In the centre of the compound was a village square surrounded by ramshackle corrugated iron huts. The Vietnamese soldiers indicated we should park our guns and vehicles in the square. We did that, posted sentries, set up our hutchies and went to sleep. Nothing to it! It started to rain. Gentle patter of drops on the metal roofs surrounding the compound. All very peaceful.

The rain continued through the night. In fact, it poured and poured, and the next morning I woke to see my webbing and rifle partially underwater! Bugger! I looked out to see the guns half submerged. Panic! I broke into a cold sweat! There was a pool of water in the centre of the compound about 1.5 metres deep. How could we have allowed that to happen? Where was the sentry? Did we post a sentry?

Sorry, Sir, we drowned the guns, was going through my head as I imagined myself speaking to the CO.

We didn't comprehend that the compound was built like a dam, with no overflow, and the heavy tropical rain had simply filled it up. The point was, we all slept through it and did nothing! In the end, all was okay. The guns dried out with no permanent damage, and the five-ton vehicles (gun tractors, as we called them) were high and dry. Soon we were on the road again. We didn't have to tell the CO anything, but we got our Senior NCOs together to sort out the night routine for subsequent nights. Another lesson for young officers.

Chinook helicopter of the RAAF delivering 105mm howitzer and ammunition underslung. This position had not yet been prepared and it was imperative that the guns were emplaced and defences prepared before nightfall to avoid a situation as happened at NDP Melissa.
Photo Commonwealth of Australia DVA

The small fire base that we were to occupy was called an 'NDP', or Night Defensive Position. From the NDP we could provide fire support for infantry patrols, which were getting more numerous in overall numbers but smaller in individual size. However, now that the actions were smaller and the patrols more widespread, the infantry battalions were unable to spare troops to help us defend these NDPs. This was clearly a gamble, as the NVA and the VC were still strong enough to attack any small gun position like ours. The intelligence people back at Nui Dat assessed that they probably would not do so. However, this meant we were now responsible for our own security, requiring serious application of the lessons we were taught only a year or so ago, in the safety of our Australian classrooms.

We only had our gun detachments plus a few command post operators and signallers to use for manning the defences. Morning and night, we would religiously send out our clearing patrols during stand-to, which are those periods just after sunrise and just before sunset, when we would change from day to night routine or vice-versa. Night routine simply meant more security, including the mounting (manning) of gun pickets and bringing on a night shift of operators to manage the command post. At night, the use of lights and torches was restricted for obvious reasons. The guns would be prepared, or 'laid and loaded', as we called it in gunner language, to engage a target selected by the FOs.

On one such deployment we arrived a little late in the day and had not fully reconnoitred the area. The NDP we were about to set up was NDP Melissa, our home for a few weeks. The helicopters that were our transport this time were late in commencing the air lift and it was almost dark by the time the last gun was choppered in. Guns and ammunition were all about the place. The absolute priority was ensuring the guns were properly surveyed into place... 'located' as we called it, and all pointing in the same direction, with the command post set up, communications with FOs established, and at least the minimum number of 105mm rounds ready to fire should there be a need to respond to any calls for fire that night.

There was no time to set up our night accommodation, so we scrambled into a defensive layout surrounding the guns. We sat in silence during the evening stand-to, all tensed up with the knowledge that many a disastrous action had occurred because of just such a late occupation of a position not fully reconnoitred. Already on edge, we were startled to hear distinct crawling and slipping and sliding noises directly to our front. Pete looked at me several times and both of us were thinking... this is it mate... we have been dropped right in the middle of a VC position! The 'gunnies' behind us were very silent and I knew what they were thinking. I hope these young blokes just out of officer kindergarten know what they are doing! Then it happened. Crash... bang! Out of the scrub burst a massive boar! A monster, then another huge pig, followed by a bunch of sows. Phew! We laughed, full of bravado. See? Trust us, we know what's going on!

That deployment position turned out to be quite an entertaining place to be. It was getting close to Christmas and there were large numbers of Christmas cakes being sent to us by the Lions, by Rotary, by teachers in schools who got entire classes of kids to send stuff, by parents and by wives and girlfriends! At first, we cheered. Wow, they still loved us! We would all gather around and eagerly share our cakes. Then the numbers arriving became serious. Bags and bags of Christmas cakes. We wondered how we were going to eat them all. We didn't. We failed miserably. At first, we used them instead of sandbags to make walls around our sleeping shelters. But this was not a good idea as it turned out—not if you want to keep the rodents and pests down!

I have no idea if pythons like Christmas cake, but they like rats! On one occasion Pete and I were lying on our stretchers, just reading books, enjoying some of that 'war is hell' time when absolutely nothing goes on. We were inside an 11 x 11 tent, so named because its dimensions are 11 feet by 11 feet.

Suddenly, Pete said in an overly quiet and controlled voice:

'Ben... I can see a snake under my stretcher, but I can't see the head.'

I froze and looked down and said:

'Pete, I can see a snake too... it's under my stretcher and I can't see the tail!'

It dawned on us simultaneously. Pete, being a big boy from the country took control and calmly said:

'On the count of three, jump!'

I needed no second invite. It turned out to be a monstrous python, substantially longer than 11 feet, because we could not see either of its ends. It was quietly crawling along underneath our stretchers whilst we were reading. Probably heading for a rat snack who knows? We sandbagged the tent with proper sandbags after that.

At that same NDP, I also managed to get stung by a scorpion. When I say 'managed' that is precisely what I mean. It took an effort on my part to do a series of stupid things that caused the scorpion to actually sting! The drill in Vietnam was to make sure you shake your boots upside down to ensure that there is nothing nasty in there before putting them on. Well, I always did that. However, I had become somewhat slack after 11 months on tour, and this time shook my boot right over my knee, whereupon a large brown scorpion, about the size of the palm of my hand, landed on my knee, hesitated a moment whilst seeming to actually fix me with its gaze, then ran up my leg and jumped up onto my shoulder in one huge leap! My instant reaction was to swipe it off, but I missed and it stung me! I should have just let it run off.

The medic had no sympathy:

'Ah, Sir! That was really dumb.'

In other words, Sir you are a dickhead!

'Sit down over there by the tree, I'll watch you. It'll be painful, but it will pass in about 3-4 hours.' He said.

Thankfully it did. A good lesson. To this day, I always shake out my boots when I am in the bush, and when I do I often think of that scorpion and avoid shaking boots over my knee. The medic explained to me that, as the colour of the scorpion was brown, he was confident that the effect of the poison was something I could deal with. Had it been a black scorpion he would have called the medivac helicopter. I don't know how painful a black scorpion sting is, but I can verify that a brown scorpion sting is very painful. I liken it to multiple bee stings, all in the one spot!

The NDP was right on the edge of a de-forested area, where the US had flown missions of pure destruction, spraying the jungle with a compound known as 'Agent Orange. The result was complete defoliation: dead trees, dead everything, no birds, no insects, nothing left alive. Surely such a thing cannot be justified. Agent Orange contained 24D and other herbicides that, as well as destroying the jungle, proved to have disastrous effects on humans. Leukaemia and other cancers were reported, not only in veterans but in their children. Fortunately, I was not affected.

Sometime just before Christmas 1970, the BC ordered all the parts of 107 Battery to concentrate back into Nui Dat. Again, the Chinooks swung into action and we left one dust bowl to head for the dust of our home base. My time in Vietnam was now coming to end and I was often asked to do extra shifts in the command post. I was happy to put in a final burst of effort, as I would be going home well before the others. The three officers who were the first of the regiment to deploy, were now the first to go home.

One of the standard brigade procedures was to conduct patrols and set ambushes to ensure that the villagers surrounding Nui Dat were complying with curfew rules. It was important to stop enemy movement at night, which is when they would position themselves for future actions. Our patrols would look up to the mountains of the Nui Dhin and see the little pin pricks of light wandering along. As the duty officer in the command post, I would then be called upon to fire artillery at the lights, which would promptly go out. They would always come back on though. It was suspected that the lights were columns of VC sympathisers carrying food and supplies. We knew that many of the sympathisers came from the local village of Hoa Long, just a few kilometres from Nui Dat. If ever there was a feeling that you were in enemy territory, it was in Hoa Long, where the mothers and the old men (there never seemed to be young men there), would look at you with a hatred that was palpable. Who is your enemy? Do you always know?

Christmas came and went and my deployment was nearly over. I was officially given my home marching orders. I had reached that magical time when I had only 13 sleeps and a 'greasy egg' to go. That was the saying: one last greasy egg from the cooks at Nui Dat and into the 'polys', the dress uniform that you last wore on R and R, that had been kept in an airtight placcy bag to prevent mould, and then the journey home. I was now on those 'happy pills' that I had watched everyone else, who had gone home before me, take. Happy pills made me feel happy, not because they cleared my body of malaria, but because they indicated a countdown of 14 days to go before RTA! All that remained was a few visits to the PX to buy the massive sound equipment that we all seemed to buy, plus watches, cameras and anything duty free. Then, before dawn on the seventh of January 1971, twelve months to the day after my arrival, I woke, dressed, and went over to Pete's stretcher and woke him up. I wanted to ensure he knew I was leaving and that he had another three months to go. Rubbing it in was standard practice. I walked out into the early light of my last day in a war zone, to the waiting Land Rover for the ride to Luscombe Field airstrip in Nui Dat. It was the first step of my journey home via Tan Shon Nhut airport in Saigon, Paya Lebar in Singapore, and Kingsford Smith in Sydney. At last I was going home to Lo and my family.

9 HOME FROM THE WAR

It took a little while to settle back into life in Australia—particularly life in the Army. Upon my return I was posted to 8/12 Medium Regiment in Holsworthy, Sydney. Starting a military career with an operational deployment first-up was not the norm in the Services, and for me it created a letdown. Being on war service first and then coming home to join a unit that was doing mundane training—in comparison to what I had just been doing—was not the best sequence of career progression. It made me restless. In any case, the brand-new Vietnam ribbons that I sported on my left breast were a source of pride to me and a source of envy to others.

I was beginning to settle in to peace-time soldiering when the CO selected me to be the Ensign for the new Queen's Banner that the corps of Artillery was about to receive. The corps still had a King's Banner, and Elizabeth II decided it was time to bestow on us her Queen's Banner. The Ensign is the young officer who carries the Regimental Colours, or Corps Banner, usually flanked two senior NCOs who are the Escorts to the Colours or Banner. In later years our daughter Raewyn, who graduated from Duntroon into the Army Transport Corps, was the Ensign for the Transport Corps Banner on an ANZAC Day parade in Sydney upon her return from active service in East Timor. We were very proud of her. Being selected as the Ensign was considered to be an honour

The parade for bestowing the Queen's Banner was held at Victoria Barracks in Sydney and involved the removal and laying to rest of the old King's Banner, and the blessing and receiving of the new one. The Queen's Banner was to be handed to me by Sir Paul Hasluck, the Governor General. Obviously, I was to be in the spot-light, so the drill Sergeant Major coached me for days and weeks until I thought my feet would stamp all the way through the floor of the drill hall! Lo thought this was hilarious as she reckoned I had bowed legs and how on earth could any right-minded CO choose an officer with bowed legs to be the centre of attention on a big parade? My sword drill was honed to perfection, or so I thought, and I was ready.

Victoria Barracks at Paddington in Sydney is a grand place. Just entering the parade ground is enough to engender a sense of place and history. Setting the scene at the rear of the ground is a 100-metre or more colonial style, two-story building with a grand, wide veranda and various archways giving access to the horse stables at the back and the offices inside. Flanking the ground on the left are stylish colonial buildings that were once the officers' quarters, and on the right, the imposing structure of the officers mess. The front of the parade ground is dotted with enormous hundred-year-old oak trees, providing a pleasant venue for spectators. Any military parade on such a ground must surely be enhanced by the atmosphere created by these impressive yet serene surroundings.

As we marched onto the parade ground I noticed there was a large crowd watching. For the first part of the parade, the King's Banner took pride of place and I was required to march with the others, my own banner party not having formed because the Queen's Banner was still to be presented. Then the moment came. The King's Banner was marched off the ground and the CO, at the top of his voice, commanded the Ensign to the Queen's Banner to come forward. This was my cue, the lead-in to the moment my place in the corps was to be cemented in history, unbeknown to me to be preserved for posterity in a large oil painting that now hangs in the Regimental Mess.

The Governor General Sir Paul Hasluck presenting the Queen's Banner to the Royal Australian Artillery. I was the ensign and received the Banner.
Photo Australian Army

Ironically, I did not see the painting until after I left the regular Army, and because the figure was too small to be recognisable, there was general disagreement as to whom it was in the painting! The corps at large seemed to have forgotten. Clearly the event was auspicious, but not the people. Quite right I suppose! Nevertheless, when I did visit the mess many years later, I claimed the moment. Unfortunately, there was no-one about that remembered, so the claim went unverified until the Army's 'Regimental Sergeant-Major Ceremonial and Protocol', a Chris Jobson, with whom I had served when we were both young soldiers in 4 Field Regiment, rang me in the late 90's. He was trying to sort out some corps history and thought that he remembered me as the young Ensign carrying the Queen's Banner on that day, and that therefore I must be the figure depicted in that painting in the 8/12 Medium Regiment Officer's Mess. I was pleased to confirm his memory of the occasion and was amazed that, after 35 years, someone had decided they should add the names of the characters depicted in that scene.

Naturally, on the parade, when the CO called for the Ensign to come forward, I was unaware that anyone should think this occasion important enough to have a picture painted and that later most officers would forget who the Ensign in that picture was!

On that day of the parade, when I heard the CO's command, I remember yelling out some sort of 'Yes... Saaaar!' in the vernacular of parade ground language, and my well-rehearsed drill movements started. Having to sheath home my sword, without looking down at the sheath itself, into the very small slit at the top of the sheath and march out exactly 21 paces to where the Governor General stood, was a challenge! Who was going to count anyway? None of the crowd would know such an irrelevancy. But it was to be 21 paces! No mistake! The RSM had made me practice again and again.

'Tradition Sir,' he said.

The 21 paces were to take me precisely to the stack of musical kettle drums—another tradition. Apparently, it is critical that regimental colours or banners, when revealed, are 'held' at all times unless 'rested' on stacked drums, or displayed in a glass case. Another piece of tradition the RSM fed me. Does this have something to do with marching into battle? Who knows?

Anyway, I made the stacked drums without mishap and had to kneel down so that the Governor General could place the end of the banner pole into a carrier cup that was suspended on a leather belt which dangled at groin height. It was a very risky place for that banner pole to be placed. Being a windy day, he had some trouble directing the pole into the right place.

'Sorry son,' he mumbled under his breath, 'I hope I don't do you an injury here!'

Luckily, he didn't. With the banner in place, I took the pole into my hands, rose, turned, and marched back to my two 'Escorts to the Banner,' ready for the 'march-past'. During the march-past, I had to lower the banner in salute every time we marched in front of the Governor General. This was easier said than done in the strong winds that day. I had practiced this move many times back at the regiment, but to my horror I found that the real banner was a great deal heavier than the practice one.

'Do not let the banner touch the ground, Sir!', came the RSM's voice ringing in my head!

We marched about for a few more laps to the accompanying music of the Eastern Command Army Band, and then it was over. Exiting the parade ground through the famous arch, which splits the magnificent colonial facade in two, I felt pretty good.

The whole of Lo's family was there to watch, along with my brother Reg and his wife Anna. They, and Lo's parents, knew what to expect of a military parade with its formality and shouting of commands. But Lo's sisters and their partners had never before been exposed to this military pomp and ceremony. At the time, I wondered what they thought about it all.

About half way through that year after I came home from Vietnam, professional boredom started to settle in. Lo and I decided on a wedding date, and that became the stand-out highlight of the year. She was still at Sydney University in the last year of her education degree. Getting married was on my mind all that year and was the reason that I asked the Artillery Directorate to post me to Sydney and not Townsville, as had been their intention. Career wise there were not a lot of exciting prospects on the immediate horizon and my professional life continued to comprise many small tasks.

One of those tasks was to assist in the firing of a 21-gun salute for Prince Philip, who was to do a small private tour of Australia. The salute was to be fired as he stepped off his VIP aircraft in Canberra. We drove from Holsworthy to Canberra in a convoy of guns escorted by police on motor cycles and in cars, with lights flashing. An armed military convoy must be given the right of way at all times. Apparently, it's the law, which not many people actually know, and why should they? We drove through intersections where indignant motorists were told to stop in their tracks and wait for the convoy to pass. The priority afforded to us was great! What a brilliant road rule!

We made it safely to RAAF Fairbairn in Canberra, where the Duke was due to land and we soon had the guns lined up, gleaming and in perfect working order, their blank rounds ready to go. Strangely, but as it turned out fortunately, the event organisers parked us so far out of the way, that absolutely nobody even knew we were there. Never mind. They would hear us! I was only the officer assisting, not the officer in charge, and my only role was to stand to attention holding my sword, count the rounds as they were fired, and stop the firing once the guns had fired 21 rounds. No problems. What could be simpler? About the only thing that might make it a little tricky was the misfire drill. This meant that, if there was a misfire, the next gun in line would fire immediately in order to reduce the chance of a gap between 'bangs' being noticed by the public.

Here we were, all dressed to the hilt with our parade-best finery on, tucked away in a paddock behind the airstrip where the only onlookers were some bemused kangaroos. I remember thinking that they would soon scatter and then there would be absolutely no eyes on us. I reminded myself again, my only job was to count to 21 and yell out 'cease fire!' Surely that could not be hard. Our boss, the Battery Commander who was on the tarmac, watched as the Prince alighted the plane. The Royal Guard saluted.

'Now' came the message over the air.

'Fire!' The Gun Position Officer yelled and 'bang' went the first gun. 'That's number one,' said I to myself. 'Bang' went the second gun! That's number two! 'Bang'... that's three! Then, click... nothing. Click again, nothing... then 'bang'! The next gun fired, in accordance with the drill!

'Aahh shit,' thought I, how many is that, three or four? 'Bang', the next gun, and I went, 'that's four or, was it five?' And so on, until the total started to approach the twenties and I knew I would soon have to make a decision. Was I one short all the way or was I correct? Forlornly I looked at the Gun Detachment Commanders whose faces were without clue. They just kept looking at me.

'We will just keep firing until you tell us to stop Sir,' they were thinking.

Okay, this is it... 21, or was it 20?

'Cease firing!' I yelled.

The guns stopped firing. The gun detachments came to attention and the gunners stood up. So far so good. Quite frankly no-one would know if we had fired 20 or 21 rounds, surely?

Wrong! The Battery Commander would know! He would have counted the rounds and what's more, he knew whose job it was to count!

'Fetch Lans, get him on the radio!' came the call. 'You idiot!' were his very next words. 'Do you realise what you have done? If anyone in the press, or indeed the Prince himself, counted the number of bangs we are in trouble. You fired only 20 rounds of a 21-gun salute!' He said in an animated voice.

I wanted to crawl into a hole in the ground but just stood there to attention, every eye of the battery was on me. They all heard the radio blare its message out loud. The seconds, then the minutes passed.

'Stand down' came the call. Quickly we got out of there. No-one ever knew, except us and our boss. Miraculously he didn't even give me any 'extra duty officers over it, so relieved was he that we hadn't been found out.

Later that year I was appointed for a stint as Guard Commander at Victoria Barracks. This was a seven-day ceremonial task, where you would conduct daily changing of the guard ceremonies for the public to view, ponce about in full ceremonial uniform to titivate the googling tourists whilst ostensibly ensuring the place was secure, and of an evening, sit around the officers mess in case some dignitary needed help finding their room. In those days before staying in hotels became the norm, senior military people and politicians would travel and stay in officers messes. The officers mess at Victoria Barracks was a grand place, and still is today. It displays the pomp and style of a bygone era One evening, I was asked to have dinner with Andrew Peacock, the then Minister for the Army, who just happened to be in town and who needed someone to chat with over dinner. So that's what I did. Polite chatting with the minister, until both of us gave up and were relieved to be able to retire to our lodgings. Many of those evenings, when I figured I would no longer be needed, I would escape and go to see Lo. It's hard to imagine the minister staying in an officers mess these days? It's all about five-star luxury now.

My time in 8/12 Regiment doesn't rate high in my career. I will admit to being somewhat slack and was frequently late, coming in from Ashfield where Lo and I had our flat, racing the traffic and melding into that day's training as unobtrusively as possible. This regiment was not going to Vietnam for some time, if ever, and there was little sense of urgency. One of the stand-out moments was when I was duty officer and I had invited Lo to spend the evening in the mess. At the end of the night it was too late for her to drive home, so I invited her to stay in my room, the duty room, which had two beds. She slept in the front bed, the one the duty officer usually slept in. The next morning the batman, who was one of the soldiers, waltzed into the room with a cup of tea, hovered above the bed, and was about to rouse the officer in bed when he realised that there was a young woman lying in the bed instead! By this time, I had jumped up from the other bed around the corner to explain the situation. I thought that Lo was about to have a cup of tea poured over her by a batman in shock!

Finally, boredom got the better of me, so I went to see my Battery Commander, who was Major Peter Sharp, my old instructor from the pre-Vietnam days at the School of Artillery, to ask him why I hadn't been given a more interesting job in the regiment, one where I could use my Vietnam experience. He obviously knew something, because he stared at me a while, as if deciding whether or not to tell me, then said,

'Come with me.'

He led me over to the CO's office. I was a bit hesitant. What ensued simply blew me away. There was in fact quite a bit of news to tell me. News that had clearly been withheld.

'Mr Lans,' the CO started, 'You have been selected for a posting to Singapore, to join the 28th ANZUK (Australia, New Zealand and United Kingdom) Brigade as the combined artillery regiment Signals Officer.

Personally, I have decided that, as you have already been fortunate to have been given one overseas posting so soon after graduation, I am resisting the Directorate of Artillery and recommending they send someone else.'

Well, I thought Peter Sharp was going to leap at the CO!

'Sir...' he began, but then thought the better of it and asked me to wait outside. Ten minutes later, out he came.

'Pack your bags Ben. You're going to Singapore.' He never did tell me exactly what was said between the CO and himself, but he did mumble something about... interfering old fool! I was incredulous. Suddenly my career was alight again. I could barely wait to go home and tell Lo. How ironic it was that we should be going to Singapore together. Ironic because Lo's father wanted her to do some travel before getting married. Well, how about this as an alternative. Let's get married first and then travel, courtesy of the Army!

Not long before that, Peter Sharp was part of the Guard of Honour at our wedding. Lo and I were married in May 1971, at St Barnaby's Church in Glebe, just opposite Sydney University and down the road from Women's Hall, where Lo was boarding these past few years. She hadn't yet finished her course at Sydney Uni, but we couldn't wait to be married. The day was arranged mostly by Lo's family, but the timing was not great, being in the middle of Lo's final university year and in the middle of my 'Gun Position Officer's' course at the School of Artillery. Irrespective of the time pressures, it was a crazy, fun wedding, with the rain pelting down outside of St Barnaby's so hard that everyone became completely soaked upon leaving the ceremony and the guard of honour had to stand *inside* the church. This was unusual to say the least, particularly for a church community accustomed to university students, who were more prone to protesting than having a military guard of honour with raised swords inside the actual church! There was no time for a honeymoon, but not to worry, we were happy in our little flat at Ashfield.

10 SINGAPORE AND MALAYSIA 1971-73

Singapore. What an exotic world. The Singapore of the Raj! This was 1971 and the world was not yet on the move. It was not yet standard practice to announce that your wedding will be held in Bangkok or Bali, and that you expect your guests from Australia, and the rest of the world, to attend. It was not yet the time for every graduate to travel the world as a matter of course and accept jobs overseas in the same manner that we once gratefully accepted jobs in Australia. It was before the time when practically every place on earth would be flooded by tourists and there would be no adventure that cannot be bought, including climbing Mount Everest!

The Singapore that I was posted to was the Singapore that was beginning to emerge from the colonial days of British rule. Where gin slings on the steps of the Cricket Club were still the order of the day. Where Australian expats were treated by the older-generation Singaporeans with a deference reminiscent of the master-servant relationship of old, and at the same time eyed suspiciously by a younger generation that was keen to step out of the shadows and into the world. The Singapore of that time engendered romantic and mysterious associations just at the mention of the name, and brought out gasps of disbelief from family members, when told we'd been posted there.

I could hardly wait to tell Lo. She would have to stay behind for a few months as I was due in Singapore in August and she still had to finish Uni, but hey... this was an opportunity of a lifetime! We had spent most of our courting time apart. It was not ideal to once again be apart, especially so soon after getting married, but in the end, it would just be a few months more. She would join me in November, after her uni exams were over. When I came home with the news she stared at me. She didn't know what to say, other than: 'Yes, yes, yes... let's go!' We were going to live in Singapore! Who would have thought? I could hardly believe it myself. It did dawn on me as the days passed that Lo really did get the raw end of this deal. We would be giving up our flat in Ashfield, our first place together, and Lo would have to live somewhere else. Luckily her sister Leonie, and Leonie's husband Richard, had the room to give her a place to stay from where she could finish her university course.

The next few months were exciting and new for me, but would be just a grind for Lo, finishing her degree. Going away again so soon after we were married was a real tearing of the hearts. We booked ourselves into Jonah's, a beautiful boutique hotel at Whale Beach on Sydney's North Shore and steeled ourselves. We bought Lo a special dress and shoes, and she swore that she would wear them when she came to join me. And she did. As I waited at the bottom of the stairs in the arrivals area of Paya Lebar, those were the first things I saw when she descended the stairs at the airport in Singapore.

However, for me two months before, flying into Singapore was a bit like flying into Vietnam, only this time I was at Paya Lebar, not Tan Son Nhut, once again looking for where the Australians are stationed. I wandered out to the taxi ranks saying,

'Okay... Australian Army, you know where the Australian Army base is?'

My posting order failed to say just what part of Singapore the Australians were garrisoned at, and that small omission had not actually occurred to me until I stepped off the plane. Naturally, all the taxi drivers knew, or so they said, but really, they didn't. They couldn't have known, because as yet there was no large group of Australians there. That was still to come. I was yet to learn that they were answering in that typical Chinese manner, where they could not admit that they did not know something, as that would cause them to lose face.

So, I found myself being taken to Nee Soon, a WWII military barracks area, used by the British since Singapore was a colony, and used by the Japanese as a POW concentration area during WWII. The driver was pretty smart to figure that Nee Soon had something to do with foreign soldiers, so that's where he took me. It turned out Nee Soon Barracks were occupied by the 1st Battalion, the Royal Highland Fusiliers, the RHF, a Scottish Regiment.

The taxi driver dropped me off at the officers mess. This was an eye opener in itself, being a stunning colonial building set high above the barracks framed with flowering frangipani trees and huge, red and green shady poinsettia trees. I wandered about with my luggage, negotiating the large entry staircase into the grand open foyer, until eventually I found the bar. Identifying the bar was not difficult in this wide, open and welcoming building, with its arched ceilings and palatial verandas. Very pucka! I struck it lucky. There at the bar, amongst a large group of Scotsmen, sat a certain Major Noel Delahunty, an Australian officer.

'Ben!' he said, 'I wasn't expecting you so soon. You have come to the right place, this is where the Australians will be, but they're not here yet. You and I are the first two!' I was astounded.

He welcomed me warmly. Major Noel, as he liked to be called, was a flamboyant fellow. Heavy smoker, drinker and party goer, always the centre of attention. A single man, who had won a Military Cross in Vietnam for carrying his wounded assistant for kilometres on his back away from a gunfight to safety and saving his life, he was quite the outgoing, sometimes brash Australian. Major Noel had me measured for tropical uniforms within a day and settled me into the mess. Almost at the same time there arrived a Lieutenant Les Mumford, Royal Artillery. An Englishman, who was posted to the Regimental Headquarters as Assistant Adjutant. Les, who was single, Major Noel, who was single, and I, who was not single, but at this stage unaccompanied, proceeded to live it up in Singapore. We were the first to arrive by months, and there was little work to do until the rest of the Australians arrived.

I was only just becoming aware of the nature of the organisation to which I was posted. The ANZUK Brigade was a tri-nation formation that was tasked with assisting the transition of power between the UK, Singapore and Malaysia, after the granting of independence. The brigade was made up of an infantry battalion from each of the three nations and an artillery regiment comprising full-time batteries from the UK and Australia, and a part-time battery from New Zealand. The artillery regiment comprised 1st Light Battery Royal Artillery 'The Blazers', and 106 Field Battery from 4th Field Regiment in Townsville. 161 Field Battery New Zealand would join us from time to time but was not stationed in Singapore. The regiment had a British CO, an Australian Second in Command (2IC) (Delahunty) and a headquarters, comprising officers from the UK, NZ and Australia. I was part of the headquarters. This was Déjà vu. I would be re-united with the 'gang' from 4th Field Regiment, the same officers that I had, as a late ring-in, beaten to Vietnam, and now, as a late ring-in yet again, beaten to Singapore. I wondered how they would take that, as they were a fairly close bunch that I had gate crashed once before.

Those first few months in Singapore were a breeze, really. There were very few soldiers with whom to do any training. It did occur to me that I could have stayed with Lo in Australia and come out with her. But the directorate had seen fit to send me early, so here I was, living in that glorious officers mess. It was one of those buildings that would be the centrepiece of any grand movie production of the Far East, a movie set representing colonial life in the latter-half of the 19th Century. The mess had large, high verandas with fluted columns and decorated arches, highly polished tiled floors and ceilings that were so high that the ceiling fans, which were two to three metres across, looked like model aeroplane propellers. A grand staircase in the centre separated a huge ballroom from the equally huge bar room, with its sweeping, long bar. At the back was a dining room capable of seating hundreds. Under nearly every arch stood an Indian steward, with red cummerbund and tray, ready to take your order, light your cigarette, and give you a chitty to sign. Cash was too vulgar; a gentleman is always good for his word, so we used these chitties that the barmen simply filled in on our behalf and we paid at the end of each month. The drinks were cheap, but in spite that, those chitties quickly became quite troublesome debts that caused a headache or two in the months before Lo came.

The first floor consisted of single officers accommodation. I had an enormous room, big enough to get lost in but with only a mosquito net-covered bed at one end and a single, tiny wardrobe at the other. There seemed to be enough room for a game of indoor bowls in between. In the style of the tropics, there were no windows in the mess at all, only louvered shutters open day and night. The grand, stylish entry foyer was preceded by large elevated porch upon which a wooden plinth supported an historic brass cannon. The cannon was of a calibre that neatly fitted a can of coke. As a result, it was frequently used to rain cans upon the married quarters across the gully—mostly empty cans, I might add. Nee Soon Barracks had a number of married quarters and those officers that were allocated quarters in the marrieds' patch situated across the gully, walked to and from the mess using concrete stairs and path built by Allied POWs in WWII. Tall shade trees, palms and frangipanis completed the garden.

When I arrived the principle tenants of the mess were the officers of 1 RHF. They were a stand-off-ish lot. They would not speak to any person before breakfast: no-one! Thus, on my first morning, all my polite 'good mornings' with accompanying grin went ignored! After a few attempts, this became quite disconcerting. I wondered if I had soiled my pants by mistake, and perhaps they felt embarrassed for me, or had I been transported to another planet and these were all aliens? But then a young RHF subaltern (another of those special Army terms meaning 'junior regimental officer') with a very sheepish look on his face, came up to me and said:

'I am sorry old chap. We do not speak before breakfast—regimental tradition.' Then he sat down and spoke no more.

Nobody else looked up. Bad manners I thought. Their regiment went back to about 1600, give or take a few decades, but that didn't give them permission to be so bloody rude! They had other strange rules! In all those hundreds of years, no woman had ever set foot in their officers mess bar, wherever that bar may have been in the world. But their dogs had! It was not unusual to be greeted by a bar full of officers, dogs at their feet, with the officers wives out on the veranda, drinking gin and tonic sent out to them by their husbands and delivered by the Indian boys. The amazing thing was that the women seemed to be quite happy with that arrangement!

Later, when 106 Field Battery, the Australian battery of the ANZUK artillery regiment arrived, the balance of mess members changed and there were enough Australian officers in the mess to force a change in some of the mess rules. At one of the monthly mess meetings, in a pre-planned and rehearsed strike, the Australian component of the mess proposed, and successfully voted in, an amendment to the mess rules that would allow women to come into the bar. Notably the British officers of 1st Light Battery 'The Blazers', that were the UK element of the ANZUK artillery regiment, were there to support us. The RHF officers were clearly upset. Us Aussies were laughing. That'll show 'em' we thought. But it was not to be. The next day we were told by Delahunty, who was now the president of the mess committee and the senior Australian officer in the mess, that the CO of the RHF had threatened him. The CO said that, if we persisted with such a proposal, he would withdraw his officers and create an international incident by informing the British High Commissioner and the British Secretary of Defence back in the UK, of the insufferable attitude of the Australians and that he could no longer work with them. Major Delahunty backed down. The women continued to drink on the veranda. That was one for the Scotsmen. We lost the fight on the mess rules but we proceeded to smash them on the sporting field and during every military competition over the next two years.

The Officer's Mess meeting, Nee Soon 1971, where the Australian and English officers
outvoted the officers of the Royal Highland Fusiliers to permit women into the bar.
Photo 1st Battery 'The Blazers' Royal Artillery

That officers mess was an amazing institution. When we had our mess do's, such as mess dining-in nights, we would sit at tables with regimental silver that went back hundreds of years. Candelabra and ornate centre pieces, presented by kings and queens, lords and ladies and other dignitaries, adorned the tables. This spectacle took our breath away every time. We became used to using, for example, silver plates and goblets with inscriptions attesting they were presented by officers back in the sixteenth or seventeenth centuries. RHF regimental pipers would pipe in the haggis, then stand in a corner of the vast mess dining room and pipe soulful highlands music. After seven or eight courses, the ladies would be invited to go and powder their noses, whilst the male officers would stay for the port and cigars. Then would follow the 'kangaroo courts' (a term used even by the British) conducted by the subalterns, where certain members of the staff, usually the more influential members on the headquarter staff like the 2IC or the Adjutant, would be accused of various crimes that they were supposed to have inflicted on the rest of us. They would always be found guilty and sentenced to several rounds of 'gunfire'... meaning shots of port. Eventually, when the men became bored with their games, they would join the ladies, who would be patiently waiting in the drawing room.

Lo was particularly incensed with this kind of male chauvinist behaviour. Quite frankly, I agreed. I thought it to be an archaic practice. The antics conducted after a dining-in night, not only in Singapore but also in Australian regiments back home, always made me feel uncomfortable because the so called 'fun' was drink-induced and forced! At times it included such juvenile games as mess rugby. It is not hard to imagine the damage that a bunch of intoxicated young 'subbies', as the young subalterns were known, egged on by the senior officers, can do to the mess furniture and glass cabinets. Later, when I was a Battery Commander in 4th Field Regiment in Townsville, Lo made her move and led a revolt by the women against the men of the regiment by refusing to leave when invited. I remember the President of the Mess Committee, who in a regiment is the 2IC, a Major, being quite astounded that a wife of one of the regiment's senior officers refused to adhere to tradition. He took it well at the time, and kind of ho-ho'd it off but he was very angry with me later.

'Hey, mate,' I countered, being of equal rank, 'she is an independent woman, and, by the way, I support her completely. Sending the women off to line up at the toilet all at the same time while the men continue drinking and smoking is, quite frankly, insulting. Tradition is great, but it's time we did away with stupid parts of tradition that have no place anymore.'

He was welcome to the traditions, and I valued some of them as well, but trying to get the ladies to leave the room and 'leave the men to it', was one tradition that Lo reckoned he had to give up. He didn't see her coming!

In Singapore, an event occurred that exemplified the behaviour of the RHF officers, particularly in the mess. I was duty officer on a night that the RHF had one of their regimental dining-in nights. The duty officer is the CO's representative, the first port of call should anything go wrong, anywhere. For example, the duty officer is the person the police might call if some of the regiment's soldiers are locked up in town for misbehaving, or if an accident happens to a soldier in the field and the family had to be informed, it would be the duty officers difficult job.

The evening was a raucous one, during which the 'coke cannon' was fired many times. This had me on edge, as I sat in my duty officers room which happened to be directly above the bar. As the night progressed, the RHF subalterns decided to get on the grog, as the saying goes. The noise from the bar increased as the hours of the morning rolled by, but that was okay. These officers of the British Army were quite used to having their own way with most things, I imagined, especially the young subalterns who, it appeared to me, were quite self-opinionated and confidently acted out the role of upper middle-class Scottish gentlemen. I readily determined, however, that the night was developing into a situation potentially out of control.

With more than a little reluctance, I decided that I'd better see what was going on. This was a disaster waiting to happen! The ruckus downstairs could evolve into uncontrolled drunkenness, and I'd be asked why I didn't intervene. On the other hand, any act of intervention by me would be akin to a colonial boy admonishing the home country aristocracy for inappropriate behaviour—a mission destined to fail.

So, I crept down and peeked into the bar. It was warfare in the trenches! Not a senior officer in sight, no-one over the rank of Captain. They had stacked the mess furniture to create barricades reminiscent of the French Revolution and were throwing wine and beer glasses into the overhead fans, apparently to replicate grenades exploding. They exploded alright. Glass after glass was hurled into the air and shattered, fragments landing at will, resulting in a layer of broken glass spread over everything. There was not a steward in sight. Obviously, the mess staff had been dismissed and the subbies had broken into the bar and gathered every glass in sight to use as ammunition. I figured I had Buckley's chance of stopping a couple of dozen or so very drunk RHF officers, so I took note of who was there and wisely retreated back to my rooms. I considered my options. Ring the CO and tell him? Not wise, I would be seen as a 'dobber'. Ring Delahunty? He would tell me not to bother him and use my best judgement to take care of it.

'Grow up Lans and do your job!' I could hear him saying! I decided to sit it out, watch and listen, and hope no serious damage to property or person would occur.

The next morning, I thought I'd be up for an uncomfortable bit of reporting, so once again I hesitatingly went downstairs, ready for that anticipated shock. But wait! The place was spotless! Not a sign of glass, all the furniture was in its place, no apparent damage. There on the bar was a note listing the damages and a cheque signed by the senior subaltern with some words annotated:

'To the duty officer. I think you'll find this cheque adequate. This is how we do things Your loyal servant, the Senior Subaltern.'

I was amazed. Here was another lesson in everything British! I suddenly felt very 'colonial' with a lot to learn in the proper behaviour stakes, right or wrong. I took the cheque to Delahunty and told him the story. He grinned. He had already been informed by the RHF CO, who in turn was briefed by the senior subaltern. They had covered all their bases. All was forgiven and forgotten.

The day-to-day work areas at Nee Soon, known as the 'regimental areas', consisted of two large squares which were surrounded by tall, three-story, traditional Asian-style colonial buildings, each floor with arched balconies. Nee Soon had once been occupied by the British Army, up to and partly during WWII, and then later by Commonwealth Forces during the Malay Insurgency of the 1950s. During most of WWII, it was occupied by the Japanese who had left it largely untouched. Many of the colonial features were clearly evident. Every day we used the path made by the POWs to take us from the regimental areas to the mess for lunch. Punjab Square, one of the two squares was occupied by the 28th ANZUK Field Regiment, the combined gunner regiment of which I was now a part, and the other square

The Blazers, the British Battery, on Punjab Square, Nee soon Barracks, Singapore,
Major Bill Hills at the front.
Photo Royal Australian Artillery Historical Company

was occupied by the RHF. During the time that we were in-barracks, that is, not out in the field conducting exercises or other forms of training, there was not much interaction between the infantry and the artillery, other than sport. What we could count on every Monday morning however, was the sound of at least one squad of 'jocks', a nickname given to Scottish soldiers, being marched down the hill past our square on the way to the lock-up, after their wild weekends out. They were a wild lot, almost impossible to understand. Small in stature, tough, nuggetty men, great when in the field, but troublesome and hard to manage when in barracks.

An occasion that stands out in my mind was a visit to the RHF square when that proud regiment was about to go back to the UK for active duty in Northern Ireland. Their replacement regiment, the Gordon Highlanders had already arrived. The two regiments were conducting their handover/ takeover. We were all invited to attend the formal farewell where, at the end of the day and the end of their tenure, the RHF 'beat the retreat', a ceremony that goes back in antiquity. This was conducted by the combined pipes and drums of the bands of the Royal Highland Fusiliers and the Gordon

Highlanders: world renowned bands of two proud regiments. Musical skill was obviously not the only pre-requisite for selection as a bandsman, as they were all impressively tall, splendid looking men. Les Mumford told me that the wartime role of bandsmen in the British Army has traditionally been stretcher bearers, so the bandsmen had to be big and strong.

To stand there, as the sun was setting over the Straits of Malacca and watch these straight-backed, imposing men in their swaying kilts, slow-march and play their pipes to the beat of the drums, was very special. On such an historic square in Singapore, where, not that long-ago, POWs were lined up and beaten... it was a memorable moment in my life! We were watching the passing of time: a world that was emerging from war and transitioning through an era of colonial domination to the dawn of an Asian age. It brought out the goose-bumps, even though none of us fully understood what we were witnessing, since we had little idea just how Singapore would thrive in the years to come.

Those first three months went by in a flash. In preparation for Lo coming over, I was allocated a married quarter in Changi. The Changi married quarter patch was near to the Sembawang Barrack area, where large numbers of Commonwealth POWs had been kept during WWII. It was also within sight of the infamous Changi Prison, from where many POWs were shipped off to die building the Burma Railway. The married patch comprised beautiful, wide, tree-lined streets, prolific with frangipani. The roads were all bordered by deep monsoon drains, not the smelly variety but purely to deal with the heavy monsoon rains. The houses were large, painted concrete structures with thick walls, smooth polished concrete floors, and windows simply furnished with metal grates, wooden shutters and no glass. This was the standard colonial style of building. Glass prevents air flow and in colonial Singapore when these houses were built, there was no air conditioning. Luckily there were no mosquitos in the patch because a little man would come around once a month and spray all the bushes. For all we knew he was spraying DDT, so we always made sure we left him to it. There was a small room at the back for the servant, with a shower and toilet, which was no more than a hole in the ground. It was still expected of every expatriate family to hire an amah, a servant who would clean, perhaps cook and, if the couple had children, look after the babies. I hired an amah called Ah Choo, simply because she arrived on the doorstep one day, having been sent by our friends, with a reputation for being a really good worker and honest. I took the chance. It was one of my better decisions.

At last in November 1971, the day came when Lo would be arriving from Sydney. By this time, I had bought an old Triumph TR4 open (soft) top. What a beast! It drove like a brick, but it felt just great... we loved it! I was lying on a public bench in the open-air reception lounge at Paya Lebar at midnight, waiting for Lo's plane. Then, there she was coming down the steps. I spotted the shoes first, then the great legs and then the pink dress! Lo! We were together again! The look on her face as we drove to our home, weaving through the markets with their steaming makan stalls, was one I will never forget. Her senses were assaulted as she first encountered the heat and the humidity, the drain smells, and the noise in the middle of night.

She instantly embraced it all and loved it! One moment she was in an air-conditioned plane in first-class comfort, and the next she was in the steamy pea soup of Asian life, in the early hours of the morning, wondering how it is that all these people could be going about their business so late at night. Later, we both came to realise that life in Singapore seemed to occur in shifts. We learned, for example, that attendance at school was in two shifts. As we drove through the throngs of people at three in the morning, it was obvious that eating and sleeping was done in shifts. There seemed to be little difference between night and day activities in Singapore life.

When she saw our house, Lo was ecstatic. She fell in love with Singapore on the spot and over the months and years this was re-enforced again and again. The taste of the street vendor's delicious wok-based food; the thick moist air that clung to your skin; the smiling and welcoming faces of the people; the colours of the orient that were everywhere; the magnificent orchids; the smartly dressed children on their way to school; the kampongs;

the peace and serenity of the married patch... she loved it all. Lo also loved our amah. Ah Choo was wonderful: she spoiled her 'Missy and Masta'. She kept that house spotless, brought orchids in every week, and washed my horrible Army gear after every exercise, picking up my clothes after me. She would wash all our clothes by stamping on them with her feet in the special large shower-come-laundry that was at the back of the house, where the floor was polished like a mirror by the feet of many amahs over the decades.

Lo had not had such an easy time of it leaving Australia. There was a mix up in her course registration that final semester and when she presented to the rooms to do her final exam, her name was not on the list! She had never been properly registered! Incredibly, within the space of time allocated to the exam, she had to rush back to the administration offices, register on the spot, and run back with her registration form before starting the paper; well behind everyone else, all hot and bothered. She didn't even know if she had passed by the time she boarded the plane, but she didn't care. She was on her way to Singapore, and to me! As it turned out she did pass, which was lucky, as going back home to Australia to sit for a supplementary exam later, was not part of the plan.

Lo managed to score a job at the ANZ High School not far from home, which was staffed by the NSW Education Department. This school was established in an old Commonwealth Forces hospital site, to teach the Australian and New Zealand children of the ANZUK Brigade. It was staffed by teachers from the New South Wales Education Department. The headmaster, who had just arrived, happened to be looking for an additional Physical Education teacher when Lo walked in, fresh from Australia and university. In no time, we were set. Good allowances, both of us working, a 'servant' in the house, an Australian dollar worth three Singapore dollars, cheap eating out at the makan stalls where the locals ate, or even at the fancy restaurants where the tourists ate, our life as newly marrieds was promising to be just fine! Lo reckons Ah Choo spoilt us forever, and she was right. Lo rarely did the ironing after Singapore!

11 SIGNALS OFFICER IN SINGAPORE

The one drawback of Singapore life was that we were living on an island smaller than the size of an average Australian capital city. From time to time we would get 'island fever', and Lo and I would simply hop into our TR4 and cross the single lane causeway into Malaysia, often on a Sunday afternoon, just for a drive. It was so simple. You flashed your ID at the border guards, who would smile and wave you across the Straits of Johore. Both our sets of parents came on visits and we took them on trips exploring the highlands and the coastal areas of the Malay Peninsula. Malaysia is rich in Commonwealth, Dutch and Portuguese history, exemplified by towns such as Malacca, that was settled by the Portuguese and the Dutch. The area is dotted with planters' clubs created by the plantation owners, or 'planters', mostly English, who created little outposts of England. We loved the clubs with their good food and cheap drinks in air-conditioned comfort. The English love their curries and in spite of their fair complexions and meat with three veg reputation, they could put away a bowl or two of fresh hot chillies! At a planter's club in Port Dickson on the west coast, I asked the waiter for some extra chilli with my curry, without even trying it. Clearly the staff were put out because the waiter returned with the hottest chilli I have ever eaten. But I was not going to be bowed. I ate every morsel of curry and chilli... and suffered accordingly! Lo and her parents just laughed.

Lo (my wife) in front of our Singapore married quarter bungalow with Triumph TR4 open top.
Photo Author

The Cameron Highlands north east of Kuala Lumpur, one of the island's traditional old colonial mountain retreats, was where in the early 70's, the indigenous tribesman wearing loin cloths, could still be seen standing by the road, holding their spears, as we wound our way through the thick rainforest that hung over the road. Here we stayed in English colonial rest houses with English rose gardens outside and tiger skins on the floor inside, reminding us that the British sahibs came here to hunt and shoot these magnificent animals. Sometimes Lo and I would go to the casino at another mountain retreat called the Genting Highlands and play at gambling with a few dollars. It was ironic that, in those days, we westerners entered freely whereas the local Chinese and Malays, who often gambled thousands of dollars, had to pay a cash surety to be allowed in.

Back in Singapore the work routine was less than arduous. The Australian Government paid for taxis to take us to and from work. I shared my taxi with Morrie Evans and Billy Foxhall, two of the 106 Battery officers. In barracks, we trained and prepared for the scheduled deployment exercises, starting at battery level through to the annual brigade exercise, when all the ANZUK units would deploy. The exercises varied in length from a week or so to about three weeks. Interspersed with that would be the support exercises, where we might go out and be part of the infantry battalions and provide artillery support, or support visiting units from Australia or the UK, who would come out for a stint of tropical training.

As Signals Officer, I trained the regimental signallers. Our big test was the brigade exercise in 1972, called Exercise Full Swing, conducted on mainland Malaysia, involving all the units of the brigade plus visiting units and ships. It was our job to establish and maintain radio nets to connect the brigade headquarters with all the participating artillery and infantry units. One glance at the map sent shivers down my back. All I could see was dark green shading indicating dense rainforest covering contour lines that were close together; the deep re-entrants, steep valleys and sharp mountain peaks of the Malay peninsula were only occasionally conquered by rudimentary roads leading to palm plantations. The country was every bit as thick as in Vietnam, but much steeper. This was going to be some challenge. Brigade decided to establish its headquarter command post in Johore Bahru, in a huge underground bunker located in a palm oil plantation. The working and sleeping accommodation was underground, dug into the rainforest floor, into slippery clay dripping with foul smelling, decaying tree litter. Excellent for the scores of little rainforest insects and animals that lived there but not suitable for us humans. All sorts of skin disorders were sure to become a daily inconvenience. Entrances were sandbagged. Fans, powered by generators that also provided all our light and radio power, were placed at random locations to blow 'fresh' air in. The headquarters command post staff, numbering around fifty personnel including our artillery headquarter cell, was cramped into this stuffy environment where temperatures hovered in the high thirties (degrees centigrade) day and night, and the humidity never seemed to vary from sticky to ultra-sticky. The feeling of claustrophobia we all felt cannot be overstated.

My command post skills were being honed and I was developing an ability to listen to several radios at once, whilst answering one and

Headquarter Troop, with me saluting the reviewing officer, on 28 ANZUK Field Regiment (Artillery) drive past on Punjab Square, Nee Soon, during a mounted regimental parade.
Photo Lo Lans

writing down information from another. Everything was sent in code, using the phonetic alphabet: alpha, bravo, charlie, delta and so on, and the transmissions were all fast, very fast. This was to minimise the time that a sender was on the air. Radio calls could be tracked or interfered with, so the shorter the transmissions the better. These days the encoding is done automatically, with the messages scrambled by the transmitters and decoded by the receivers. The technique of frequency hopping is used, where radios automatically hop from one frequency to another and use a dozen or so frequencies to transmit a message of just a few seconds. But in those days, it was all done manually, in code, on one frequency. The codes were held by the brigade signals staff who released a new pack every twelve hours. The release time was based on Greenwich Mean Time (GMT), so there could be no confusion. The codes were designed to be used throughout the world, facilitating secure communications between all NATO and allied countries. Old packs of codes were burned under supervision.

Artillery fire plans consist of many lines of detailed target and timing information and I became skilled at coding and decoding them in a short time, whilst still aware of the radio traffic on the other radio nets. It is amazing how the brain can be trained to listen to a great volume of information and decipher what's important and what's not, amongst the distractions of a command post filled with people. It is a skill which has stayed with me, being able to distinguish between and acknowledge a number of conversations at the same time. When I need to, that is.

Plantations were prevalent with snakes: large king cobras. Snakes keep down the rats and other vermin that do a lot of damage to the trees, so the plantation owners encouraged the snakes. They were everywhere. A bite was undoubtedly serious and a spit in the eye, which is what a cobra often does prior to attacking a prey, is blinding—at least temporarily. Because there was limited space in the headquarters underground sleeping area, those of us who were duty officers in the command post, used a hot bed system. That meant that you shared your sleeping space with another, who would use your space to sleep during the time you were on duty. The underground sleeping spaces were not much more than a body-sized scrape into the wall of a tunnel. They were even more claustrophobic and lacking in air than the command post itself. One night, the artillery duty officer who was finishing his shift, came down into my part of the hole in the ground to shake me awake.

'Okay mate, I'll be there in a few minutes,' I said. I jumped into my gear and trudged up to take over.

After handing over, a process that took about ten minutes, the officer, a Scottish Captain, went down to sleep on the stretcher that I had vacated. Within a few moments, there he was again... visibly shaking!

'Mate, what's up?' I asked. In his Scottish way and with a pasty white face he blurted out:

'Bloody hell, you seen what's in your bed? What were you sleeping with for Christ's sake? Your pillow! The jumper you were using as a pillow! It was full of lumps, mate, so I punched it a bit, put my head down, and the damn thing moved! Cranky King Cobra leapt up, didn't he? Shit! Come and look!'

We both rushed down, leaving the radio monitoring to the signallers. The cobra was nowhere to be seen, but we did see the hole in the side of the overhead revetment that the snake must have used to get in and out. Revetment is corigated iron sheeting, used to support the sides and ceilings of bunkers. Apparently, the cobra must have entered after I left the bed and before the Scotsman jumped in. It probably thought that my pillow jumper was a nice warm spot! Alternatively, the cobra had slid under my head whilst I was asleep! That didn't bear thinking about but was a real possibility. It had happened to other soldiers in Malaysia, where someone had woken up to find a snake inside their sleeping bag with them, all curled up and warm. A snake trapped in a sleeping bag with a person would probably find it hard to rise up and strike because I don't remember anyone actually being bitten in such a circumstance. The snake wouldn't need to bite. I reckon you'd die of fright!

During that exercise, communications were difficult to maintain. The units of the brigade were all out somewhere in the 'ulu', as we called the jungle, mostly unable to communicate with brigade headquarters. Successfully sending a radio signal from deep inside the rainforest where the unit patrols were, to the headquarters which was surrounded by almost impenetrable palm-oil trees, was ambitious. Gunners had a reputation always getting through, so to prove the reputation was not unwarranted, I went on a reconnaissance with my little troop to see what could be done. We found a small clearing about 100 metres from the headquarters, which had a conveniently-placed mound, right in the middle. This mound presented us with a platform about three or four metres above the plantation floor, just high enough to place our radio masts and allow our antennas to poke up above the canopy! We dragged our antennas and the cables to the clearing and... hurrah! We achieved excellent comms with our units and FOs. For a day we boasted the only reliable communications network of the whole brigade, until the other brigade communicators copied us and set up their equipment next to ours. Still, we basked in the knowledge that once again the gunners had led the way. A reputation maintained!

As the exercise progressed, units moved beyond normal comms range and our little mound was not high enough to keep the lines open. Thick jungle swallows radio waves and once again, my little troop needed to exercise some initiative and re-establish radio comms. So, we had to find another way. I decided to take the boys out to find an accessible peak, where we could place our receiver/transmitters to act as a relay station, or, as we called it, a 're-broadcast site'. There was just one problem. The site would need careful monitoring. Batteries would need changing and daily frequency changes would have to be made, all manually. My troop was an interesting mix of Pommie and Aussie soldiers, who got on with each other remarkably well, but they were not jungle-trained, and tended to look to me for guidance on all things to do with living in the jungle. So, when I said, with a gnawing sense of self-doubt, that whoever was left on the hilltop to maintain any re-broadcast site that we might establish would be just fine, they all accepted my word without question.

I had previously used a helicopter in an attempt to find a suitable mountain with some space on top, where the trees thinned out and there was sufficient room to establish a bunch of antennas and a few hutchies for my men. The pilot tried his best to work his way up the sides of mountains that I had identified on my map, but the monsoon rains were getting ready to unleash their annual watery onslaught and the cloud cover was too low, bringing visibility down to just a few metres and making safe helicopter flight impossible. Several hair-raising attempts to avoid trees and follow the contours nearly ended in disaster, when the blades came close to kissing a few branches. So, we abandoned that idea. Now, in a small convoy of Land Rovers, we set of on a prayer and a hunch. The map indicated no tracks leading to the tops of any mountain at all. Nevertheless, I felt confident that some of the forest workers would have made at least some tracks that might lead to the tops of the mountains. If not, we would have to hack our way up.

My team comprised two Aussie and two Pommie soldiers with an Australian NCO as leader. We would use Land Rovers to establish them in location.

They were to be issued with live ammunition, as the risk posed by wild animals was real. For two days, we bumped and slid our way up winding slippery tracks in our Land Rovers, going at not much more than walking pace, the vehicles in four-wheel drive, low ratio, 2nd gear. All the tracks were dead ends. I tried to read the map, but this was difficult in dense, steep terrain which had few reference points. Finally, after a two day search, one track petered out into a faint trail which looked more like a tunnel winding up the side of the steep hill. Promising.

We were still hundreds of metres from the summit, but the vehicles would carry us no longer. It was getting dark. One of the drivers spotted another tunnel track crossing at right angles. We could now park the vehicles and camp the night and turn them around for the descent. That was a relief as I was wondering for some time how we might achieve this feat. We used the cross track and lined the vehicles nose to tail then broke out the rations packs, made hot brews and cooked up some good old bully beef. No matter how long the Armies of the world have been making patrol rations, and how much science has gone into designing a nutritious diet, bully beef always seems to remain part of the menu.

Jungle sounds penetrated the evening peace. Woops and whistles and howls...I'd been in a lot of deep jungles in my various deployments, but this spot was about as dark and deep as anything I'd seen. We arranged our sleeping bags underneath each vehicle, because that was about the only available space, and settled in for the night to the incessant noise of moisture dripping all around us and the unnerving sound of monkeys woop-wooping over our heads. It seemed there was nowhere in that forest where you could escape the monkeys.

The first few hours were uneventful, then all hell broke loose! Waking suddenly, it took a moment to understand what was happening... but there was no mistaking what was making that noise. It was an elephant, furiously smashing its tusks against the end Land Rover, and intermittently roaring and trumpeting and rearing on his heels, only to come down with a mighty stomp on the vehicle, with the hapless crew forced to lie underneath—terrified—hoping that an elephant was not strong enough to move a Land Rover. Thankfully the elephants of Malaysia, although big enough when you are lying on the ground, are not as big as the Indian ones, and the Land Rovers stood their ground! Eventually, the crashing and bashing stopped. The elephant took to the vegetation along the side of the track and, after crunching and smashing its way loudly paralleling the convoy, continued in the direction it was planning to go, still angrily trumpeting. Was there going to be another one? We lay still for a while; nobody game enough to see if the angry elephant had moved on. Not much was said. There was not much to say really. But nobody slept after that.

As the light of a new day started to break the canopy, we emerged and we all quickly realised what had happened. We had actually parked the vehicles nose to tail along a tunnel-like track that was an elephant path. How clear it all seemed in the day light! Perhaps we had managed to upset this particular elephant, which was probably using the track to go home to the missus, who knows?

From here the communications team was despatched, looking a little less confident than they were the night before. I did worry as I watched them make their way on foot further up the hill to establish the re-broadcast site. The monkeys in the trees overhead took up their woop-wooping again as they followed the men, who looked like they were heading off to meet with Dr Livingstone or some other long lost explorer. I waved cheerily and smiled, secretly hoping I was doing the right thing. This place was so remote and so thick that it would not be out of the question for these men to get lost. There was no Global Positioning System in those days. If something happened to them it would be my responsibility. Had I done enough checks? Did I brief them sufficiently? Was I asking them to do more than they were capable? I checked my map a dozen times to make sure I could find this spot again, and we watched them disappear into the dense green never-never. But I had faith in the Bombardier I placed in charge.

A half day later we heard from them. 'Loud and clear over'. Success! We'd done it again. They were at that site for a week. Fortunately, the top of that small mountain had a partially cleared area that, with a little work, became an excellent communication site. The site was big enough for them to hack a heli landing pad for heli extraction. The men had a ball. The two British soldiers hadn't seen this kind of terrain before and although the two Aussies had a little more jungle experience, they all regaled in their stories of jungle sounds and experiences when they came back.

12 Forward Observer Officer in Singapore

In 1972, the newly promoted Lieutenant Colonel Noel DelaHunty replaced Lieutenant Colonel Trefor Jones as CO. There was a marked difference of appearance between the crisp well-dressed Briton and the casual Australian. DelaHunty was not the most 'regimental' officer, managing to look permanently crumpled in his uniform even in a place like Singapore where all the laundry tasks were performed daily by an Amah. He also sweated profusely all the time, always dobbing his brow with a hankie. Clearly that didn't help. He was known for his gravelly voice, which made it difficult to understand his words of command on a formal occasion such as a mounted parade. In all, he was the complete anthesis to the smartly dressed, clipped-voiced British CO whom he replaced.

This is how some members reflected on the first mounted parade conducted by the regiment upon the appointment of the new CO.

On the first order the assembled troops would come to attention and officers would salute. On the 'mount' order the officers would come down from the salute, and all would smartly move to their vehicles with officers and Gun Sergeants standing and left arm braced on the windshield. On 'start up' the drivers would start their engines and when each vehicle was turning over the vehicle commander would raise his right arm to signify the vehicle was ready to move. On the order 'drive past' the vehicles, by batteries, would move forward in turn. It was a delight to behold.

The first parade with the new commanding officer in January 1973 things did not go according to Hoyle. The troops, vehicles and guns were positioned by the Regimental Sergeant Major; the officers posted by the Adjutant, the parade handed over to the Second-in Command and finally to Lieutenant Colonel Delahunty, lord of us all. The Colonel took a deep breath and roared out 'RRREGIMENT ARRRGH'.

1st Light Battery RA came to attention, half the officers saluted, the other half climbed into their vehicles. 106th Field Battery RAA did not move at all, at least not until the Battery Commander saluted then gazed around-saw that some Brits had climbed into their vehicles and did likewise. Half the battery followed. Half of Headquarter Battery saluted, climbed in their vehicles, started them and began to move. The CO stared in disbelief at the events he had put in train, about turned and marched off the parade ground.

A few minutes of general milling occurred before the Second-in-Command plucked up the courage to scamper of the parade ground to seek further guidance. He reappeared from the headquarter building to call for the Adjutant who scurried off merely to quickly appear again and order 'Fall Out the Officers'! We gathered behind the building to be told that regardless of what the CO appeared to order on any given parade we were to come to attention... pause... salute... pause... mount up... start up... pause... wait for the CO to mount his vehicle, then drive off in order. We did exactly that for the next year.'[1]

---

1   The Forgotten Regiment, 28 ANZUK Field Regiment, Royal Australian Artillery Historical Company

The six 105mm L5 Howitzers of the Australian 106 Field Battery, on a mounted parade on Punjab Square.
Photo Royal Australian Artillery Historical Company

Noel Delahunty immediately set about moving his officers around, to give them different experiences in their professional development. I was seconded to the British 1st Light Battery—'The Blazers'—as a forward observer, and given my own small team comprising a forward observer assistant, known as an 'FO ack', and some signallers. Now part of a British battery, my men were British soldiers.

The Blazers were a British unit with a history that Australian units only dreamed about. It was raised in 1779 and saw service in such places as the Crimea and India. In World War 1 including the Somme and Ypres, the battery position was gassed during the German Spring Offensive of 1918.

> On 27th May 1918, the Battery suffered its worst casualties of the War when five guns in action at Bois de Mines became surrounded and although the order to remove breech blocks was given, none made it back, and only eight survivors reached the wagon lines. The Battery withdrew for the rest of May and thereafter saw no further action till October and the final Allied Advance to Victory. The last day in action for the Battery in the Great War was 27th October 1918.'[2]

2   The Forgotten Regiment, 28 ANZUK Field Regiment, Royal Australian Artillery Historical Company

Here I was... an Australian. I was suddenly part of all this history and tradition.

The training exercises that we went on in Malaysia were either 'live' or 'dry' exercises. Live firing could only be practiced on a live firing range, where high explosive ammunition could be used under controlled, safe conditions. Dry exercises meant there was no live firing of the guns. The guns would be deployed into the jungle often by helicopter. This was old hat for the Australian gunners of 106 Battery but a vey new experience for the British gunners of 1st Battery. The helicopters that were available to the regiment were the Wessex 'Whirlwind' helicopters, affectionately referred to as 'the flying boot' by the gunners. They were of an earlier generation than the Hueys that the Australians had been using in Vietnam but they did their job well and made just as much if not more chaos with the downwind from their blades.

A number of times per year the gun line and the FOs needed to train with live ammunition. In my case, as an FO, I needed practice in adjusting the 'fall of shot' onto the target. There were two places where the battery could conduct live artillery practices. The first was a tiny artillery range that the Singapore Army had developed, and the second was a larger artillery range in Malaysia, that was developed by the Brits but was now managed by the Malaysian Army. The one in Singapore had the guns firing from one island to another, with the Observation Post (OP), comprising the FOs and their signallers, situated on a lighthouse located on a third island. The firing could only occur at low tide, lest the guns become submerged in the rising waters, because the island containing the guns disappeared under water every high tide! The logistics of using this range were significant, particularly considering helicopters were needed every six hours or so to move the guns. As a result, we rarely used the Singaporean range, but it was great fun when we did!

As a new FO with 1st Light Battery, I experienced a different kind of leadership under a less than conventional commander. Major Bill Hills, the British Battery Commander, was not your everyday officer, especially when compared to the somewhat less flamboyant Australian Battery Commanders I had served under to date. He had a professional, yet on the surface, cavalier approach to most things, believing that training should be serious but also enjoyable! I learned a lot from 'Major Bill' as he was affectionately known. He decided that, in between the tides, when we could not fire the guns, the FOs would entertain themselves by playing bridge and drinking beer. Every time the tide came in and the guns were airlifted to shore, the friendly lighthouse keeper would be dispatched to the mainland in his motorised sampan to bring back plenty of ice cold Tiger beer. This was exercising in style! I was getting used to working with the Poms!

The other live firing range was at Asahan, Malaysia, about two hours' drive north of Johore Bahru. Asahan was fun! It was a conventional live firing range, full of unexploded rounds as well-established and long-time artillery ranges tend to be, and it had a large base camp. We would do our live firing deployments and at night time all the gunners and the FOs would congregate at the camp for hot showers, fresh food and cold beer!

Other exercises were about jungle patrolling activities, conducted in the central highlands of mainland Malaysia, in hot tropical jungles where the tigers and elephants still roamed, and monkeys abounded. Being on patrol as FOs, in support of our infantry mates, I found that the jungle was so dense and impenetrable that my eyes became accustomed to focussing at the short distance the eye could see and would adjust rather painfully to long distances on the rare occasions where the forest opened up. Communist Terrorists, locally referred to as CTs, also frequented these jungles. In some of those remote areas, they were left-over dissidents from the days of the communist Malayan insurrection that occurred in the late fifties and early sixties. On a few occasions after the ANZUK brigade had conducted an exercise, the CTs would march down the main street of the local village to demonstrate that they were not intimidated by these Commonwealth forces.

I was learning to manage my team of British soldiers. They were different and they all, particularly the Bombardier, regarded me as distinctly 'a class above'. Clearly, the class system was not dead in the British Army. Officers and 'other ranks' were still divided by class. They thought it quaint to have an Australian officer as their boss, not sure if he was any good, but they were polite, and I could see they were observing and judging me. On our first big live firing exercise at Asahan as a new FO team, we were tasked by the Battery Commander to dig an observation post (OP) into the hillside and adjust the fire from there.

'Okay boys, this is the spot, let's do it,' I said, as I took out my shovel and started digging. I dug for a bit, then realised that I was the only one digging! There they were, the others... all staring at me!

'What's the matter?' I asked.

'Sir, you're an officer and we're your team. Please put down that shovel and let us dig the hole for you,' my Bombardier said.

I was astounded. I said, 'Okay fellas, let's get this straight. I am an Aussie officer, not one of your Pommie officers. If we're digging a hole, then I will dig that hole with you, unless I am required to do something else. We're in this together.'

The first lesson in international soldier-to-soldier relations had just happened.

The Bombardier in particular was very wary of me for a long time. He was an experienced old hand and he was not sure this colonial officer knew what he was doing. But I must have passed his assessments and before long we became a closely-knit team. Which was a good thing, as the members of an FO party tend to spend a lot of time uncomfortably close to each other... literally, holed up in unpleasant OPs staring through binoculars or typically sleeping cuddled together in rough terrain under one hutchie, because there was not enough time to establish proper sleeping conditions.

Being an FO meant that not only did I go out on the artillery exercises, but when the Kiwi and Aussie infantry battalions went on exercise, they would need their FO party along, so I would go on their training exercises as well. These were always on mainland Malaysia and always in the deep rainforest, because that is what we trained for—counter insurgency warfare in close country.

The Aussies and the Kiwis were experts in jungle patrolling, learned from a long history of campaigns in the jungles of New Guinea, Indonesia, Malaysia and Vietnam. In the close terrain of the jungle, where visibility was limited to just a few metres, correct spacing between individual soldiers, platoons and companies was critical. Forward scouts moved slowly and silently in front of the forward platoon, up to 30 or 40 metres ahead, always under the cover of the following rifles. My Bombardier and one of my signallers were always up with the leading platoon, and it was their job to update me and immediately call for artillery support if needed. In Vietnam, this was for real. Here in Malaysia we were training, but we took it just as seriously. 'You fight only as well as you train' is the saying. Naturally, all this deliberate movement made progress very slow, which was just as well because we FOs had to map read and be deadly accurate in knowing where we were all the time. This meant continually studying the map, counting paces, and

An L5 105mm Howitzer being deployed into a battery position by a British Air Force Wessex helicopter.
Photo Royal Australian Artillery Historical Company

noting the direction of travel on our compass.

As a Lieutenant or Captain FO, I was always with the company commander, who was a Major. I listened to his orders and planned my targets along the way, in case we encountered the 'enemy'. Towards last light, the company commander would call in his platoon commanders to give them orders for the night. This usually involved one of the platoons setting up an ambush on a potential approach to the night position, and the other two preparing 'harbour' positions, with company headquarters somewhere in the middle. In each platoon, machine guns would be deployed at twelve, four and eight o'clock, with twelve o'clock being the direction of travel, and with each machine gun situated so that it could provide covering fire for the other machine guns.

The section commanders then indicated to their sections, where the weapon pits were to be placed, with two or three pits up front and a couple of pits 'in depth', each pit always covered by another. Depending on the level of threat and the length of time that the position was to be occupied, each weapon pit was dug-in, at least a foot or so deep, to help avoid injury during a mortar attack. Temporary tracks from each personal weapon pit to the gun pits were marked around the platoon area, using vines or string, and sometimes telephone land-line was laid. This was to allow the platoon headquarters to communicate with their forward pits and vice versa, and with company headquarters without using a radio.

A bit of personal admin—like eating and brewing a hot drink, rolling out the sleeping gear and setting up a low hutchie for the almost-certain nightly tropical rain—filled the time until stand-to, which was when the whole defensive position would fall silent, each person with their webbing on and weapons at the ready. Historically, sunset and sunrise are times when attacks occurred. Clearing patrols were deployed to circle the area as a last assurance that we were not being probed, sentries were posted, and as the sun sank, night routine began. Those that were not on picket duty slipped into their hutchie space for some sleep, always in the same sticky, sweaty and smelly jungle greens that had been sticking to their bodies for days! No PJ's out here! Most of us slept in sleeping bag liners—it was much too hot for a sleeping bag! Night routine meant no lights, no noise, lying on the rough, invariably soggy rainforest floor, usually crisscrossed by tree roots and ant trails, and being generally uncomfortable. Sweat would be beading in your hair, trickling down past your eyes, and dripping off your nose, until you could stand it no longer and wiped it off, only to accidentally rub mosquito repellent into your eyes.

Each machine gun pit was manned by a double picket, that is, two patrol members at a time. The tracks that were prepared earlier enabled each soldier to get to the machine gun pit and back to his hutchie in the black of a jungle night without getting lost or potentially getting shot by the sentry. One of my class mates from Portsea was killed that way in Vietnam, shot by his own patrol members, after coming back into his patrol base from the wrong direction. In the morning before first light, the clearing patrols went out again, and after a quick breakfast of hard rations, the position was cleaned up, stores packed, and we were on the move again.

One night, when we were holed up, occupying a defensive position in deep jungle, bedded down for the night and secure in our hutchies with sentries out, a chilling sound made everybody freeze:

Why am I awake? What's that noise? It's dark... really dark! Black. I'm all sticky! I'm dripping! Where's my hutchie? Ah there it is, just above my head. Bloody hell it's dark! That's right, I made a stretcher to get up off the ground. But the hutchie wasn't tall enough so my face was sort of pressed against the hutchie all night. There's that noise again, a grumble, a purr? Wow, really deep purr! But not loud. Tiger? Shit! Am I the only one awake? There it is again. It's going away. Phew!

Uh oh, it's closer. That animal is circling us. Whoooa...! Must be a tiger! What else? Hang on, someone's whispering. Thank goodness! Others are awake too. Who's got the weapon with live rounds? Was it the Platoon Sergeant? Yeah, I think so. But where's he? Would he shoot it? He probably can't even see it! I hope it's gone. Can't hear it now. Take him not me! I'll just lie still!

In the morning, we all talked about it, but we were never sure. Tigers still occurred in this part of the world, so it was entirely possible. I doubt if a soldier with a rifle would have been any match for such a superb hunter in its own environment. We were, after all in the domain of the tiger. Funny thing is, the sentries, who were wide-eyed and had also heard the gentle sounds of the circling tiger, had not been able to spot it. I think it was a tiger that found us. It wouldn't be hard for a wild animal to find a large group of smelly humans in that environment.

The rainforest was so dense and full of thick woody vines, that we carried secateurs in our hands. Popular belief has every jungle patrol hacking its way through the undergrowth by smashing a path with a machete. Not only is this incredibly loud and would warn any enemy of an approaching patrol, it is unbelievably tiring. Far easier to carry sharp secateurs and gently snip away the offending vines and other bits and pieces. Our favourite was a vine called 'wait-a-while'. It has long tendrils that hang down from the canopy, with spikey barbs sloping back in such a manner that, if you brush up against the plant, it will dig its barbs into your clothes, your pack or your skin, and hold you there. There is no getting away. It can open you up like a zipper. You simply have to stop, take a few steps back, extricate the barbs, and move on, knowing that a tiny little point of the barb will have broken off and be sitting there under your skin, waiting to fester in a few days' time. Hence the name 'wait-a-while'... wait and be patient before you force your way ahead, then wait for the pain, and then wait for the infection!

Major Bill tasked me to act as 'enemy' Platoon Commander for a training exercise designed to familiarise a troop of British Royal Marines in conducting patrols in a tropical jungle environment. The unit was 40 Commando, a battalion of the 3rd Commando Brigade, stationed in Cyprus. The Royal Marines were formed during the reign of King Charles II in 1664, and in the ensuing three and a half centuries served in more battles on land and sea than any other branch of the British Armed Forces. Thus they commanded an immense amount of respect and had an air about them that said: 'don't mess with us'. They were amongst the best of the British forces, but these particular Marines knew little about rainforests. As Aussies, we had the most recent jungle experience, so the BC asked me to manage the task of showing them the ropes.

Teaching Pommie gunners how to act like a group of revolutionaries hiding in the jungle was a lot of fun. We set up a camp replicating the typical villages I had seen in Vietnam, with booby traps, attap huts and tunnel entrances that hid explosives and ammunition. We didn't actually dig any of the tunnels as that was just too much hard work in such a short time, but we created entrances that simulated tunnels. In the centre of the camp was a small attap-covered meeting place, with a flagpole which had someone's smelly underpants mounted at the top. I can't remember seeing anything like that in Vietnam, but it seemed a nice touch! The meeting place was where we ate, discussed our plans and plotted to conduct interrogation of any unfortunate Marine that we might capture. Surrounding the camp, we set up smoke grenades to replicate claymore mines attached to razor-thin tripwires. Knowing that the Marines were still a few days away, I allowed the boys to build a huge fire next to the meeting place, and we sat around at night telling stories and laughing at dumb Pommie and Aussie jokes, not to mention having the odd nip of rum. A bit of alcohol works well to warm up international relations! To put the finishing touch on the position, I established an ambush on the most likely approach. The ambush position was carefully selected to ensure that the Marines would have to fight their way through an impenetrable thicket of wait-a-while. We did have the advantage of knowing they were coming from this direction, as the exercise had been scripted that way.

The ambush was sprung in daylight. The Marines, as per their training, immediately turned and fought their way into the oncoming fire. But, they had not anticipated this crazy plant called wait-a-while! Never before had they seen anything like this! I will never forget the look on a young Marine's face who, when he saw me, charged me with rifle to the fore, intent on drilling me between the eyes, but then stopped dead! There he hung, somewhat inelegantly suspended by the wait-a-while. Being a tough Marine, and not understanding the nature of the plant that was holding him, he refused to stop trying in vain to fight his way through, until the spikey vines enveloped his entire body, pack and rifle and he hung there, perplexed, throwing his arms and legs about, staring at me with desperation and disbelief, not even able to bring his weapon to bear. A quick signal from me and the gunners turned and slipped away, leaving confused, suspended Marines everywhere. It was a hard, painful lesson for them, but I was absolutely sure that not one of them would ever forget it. Particularly not for the next few weeks, as the little tips of the barbs festered in their skin. It occurred to me that I had better not get caught by these blokes. They may take it out on me with some over the top interrogation training!

After the ambush, we withdrew to our hilltop compound to await their approach. It was here that we went into normal routine, acting like a guerrilla force, quietly living in the forest, not attracting any attention through undue noise. The compound was tactically sited, just over the crest of a large hill, commanding good views over the approaches, as good as they could be in the limited visibility of such heavily timbered terrain. It took several more days for them to find us and when they did, there was the expected battle and, naturally, we were overrun by a superior force. They tied us up and bundled us in groups, heads covered with bags, ready for some aggressive questioning. Fortunately, the BC stepped in and the exercise was halted at that point. It had all been great fun though and everyone had a good laugh about the wait-a-while. Well, the gunners did anyway! The Marines thought it was about as funny as we had anticipated.

The Marines had their own doctor with them. He was a fit young man, trained to treat battle injuries and apply that first level of medical support. He had travelled all over the world with this Commando unit. However, he had no jungle experience. He spent his time with me and my men instead of being with the Marines so I tried to teach him some simple rules of life in a thick jungle, such as staying away from the streams, tempting though they may be. Streams in deep tropical rainforest have a habit of presenting unsuspecting surprises, including a wide variety of swimming or floating creepy crawlies, such as snakes and just about every other animal in the jungle that needs water. The doc liked to dangle his feet in streams. I warned him several times.

'Mate, you have no idea what nasty things may be eyeing off your toes, not to mention the millions of unseen little bugs looking to crawl into every little break in your skin, infect every small cut you may not even be aware of,' I said.

But he couldn't resist. One day, having washed his feet in a pretty little stream, he pulled out his foot, to discover a stripy 'tiger leach' attached to one of his toes. Not just any leach... this was a river monster! The leach was so big that its suction mouth covered the *entire* top of the second toe! It took a burning match to make it let go. He never did that again!

I think I managed to teach them a few more tricks to help them survive in a tropical jungle. For example, never step over a log or other object without knowing what is on the other side, unless you don't mind stepping on a snake. Oh, and don't hang on to branches and vines as you push your way through the thickets, or you may pull down all manner of bugs, beetles, ants, spiders and more snakes, straight down the back of your shirt! This is very inconvenient as your shirt is bound to be soaking wet with sweat, and will not come off easily, not least because it is under your webbing and probably your pack as well. *Never* drink stagnant water, for fear of getting leptospirosis, which one of our officers did on another exercise. It nearly costs him his life. From the leptospirosis, he developed meningitis and lay in a darkened room in Changi Hospital for many weeks, near death.

I also taught them to sleep off the ground, whenever that was tactically possible, on a stretcher made from saplings tied together in a cross at each end and using the ground sheet (the one that the Army designed to contain those blow-ups that we used at Portsea all those years before) suspended by means of two long poles fed through the outer slots. This creates a stretcher which helps to keep you above the jungle floor and away from its daily traffic, that may include all or a variety of centipedes, scorpions, leaches and an endless trail of every kind of ant you have ever imagined! I taught them that leeches were our everyday companions, fabled to find warm, welcome skin to attack everywhere. Sometimes, when on patrol, soldiers would sit back to back flicking approaching leeches away.

One night, I must have been dreaming, because I apparently sat up in my stretcher and yelled out loud... 'Doc, Doc...watch out behind you!'

The poor old Doc nearly jumped out of his skin and was in quite a panic as he screamed at me:

'What? What? Where?' He shook me.

'What do you mean?' I said. 'What's up with you?'

After the experiences with the leech and other creepy crawlies, the Doc was on high alert for everything and anything in that jungle!

During one exercise, we ventured for several days into a dark and murky mangrove area, where we had to jump from tree root to tree root to stay above the swampy water, and at night we had to suspend ground sheets between the tree roots to remain dry. Juggling map, compass and rifle, whilst gingerly avoiding the black slimy water, I momentarily lost balance and in the ensuing ungainly grope for a tree branch, I lost my compass! It slipped out of my hand. My beautiful, brass, Army-issue prismatic compass, almost an antique... plop, bubble, bubble—down it went into the inky depths! For a moment, my emotions got the upper hand and almost succeeded in forcing my brain to send my hand after the compass. However, common sense won, and I let the thought go. I stared at the bubbles instead. That compass will still be lying there today, abandoned at the bottom of a black, watery grave.

Dangers associated with the dense nature of the forest were demonstrated another time when one of the Kiwi officers was accidentally killed when hit on the head by a sandbag full of rations during a standard food drop. Normal procedure for food resupply was to use helicopters to drop ration packs and jerry cans of water at agreed rendezvous points. Often there was no clearing large enough to land, so the helicopter would drop the supplies to waiting troops below, through the jungle canopy. On one notable occasion, it all went wrong. A sandbag full of ration packs crashed through the canopy and struck a young officer on the head killing him instantly. When the news reached Changi, Lo was visited by the Army Padre. This particular Padre was something of an odd fellow, with little compassion—not the typical sort of Army religious man. He told Lo that an officer, who lived a few doors down, had been killed and to prepare herself to assist his wife. With that he charged off and moments later she heard the anguished moan of the woman just up the street. Lo frequently recalls that event and wonders how he broke the news to her, or how Lo herself would have reacted if the Padre had delivered the news that I had been killed.

Being in Singapore had many benefits. We had a lot of spare time on our hands, and that was spent playing sport. Sport was a big deal. The brigade units all participated in football competitions. Rugby union was big with the Aussies and the Kiwis, and I was occasionally asked to play for the 106 Battery team, mainly because I was an Australian Rules footballer by background who could kick and catch a ball. Thus, I was placed at fullback on those few occasions when the union team was short of players. The most memorable game was when we played a demonstration match on the famous Padang. The Padang is the hallowed ground just in front of the Singapore Cricket Club in central Singapore where, during WWII, POWs were lined up and marched off to prison camps like Changi or put on a train to Burma to work and most likely die. On the Padang, they were lined up, usually without food or water, and in some cases, executed. Now that hallowed ground was used for sports matches. I could not help but feel the heavy significance of the place.

The football I lived for however, was Aussie Rules, later to become known as AFL. The Aussie Rules competition was a vibrant one, comprising teams from all the Australian units. Each Saturday, at various ovals around the barracks areas, cars would surround an oval in the same manner they do in Australia. Footy teams, dressed in the colours of the Victorian Football

League (VFL) teams, would run onto the ground, much to the amusement of the local Singaporeans who neither understood nor cared for the game. 106 Battery team played in Essendon Colours. Years later, when my son, Matt, was old enough to begin to understand football, he saw my footy jumper and reckoned that I must have played for Essendon. He told all his mates and for a short while I was a footy star!

Athletics was also a big deal, and every year the units would stop all other training and prepare for the athletics carnival. 1st Light Battery was particularly keen to win the Tug of War. This was a prestigious event in the UK and Europe, where 1st Light Battery was the reigning BAOR (British Army of the Rhine) Open Champion. This was no small thing, as BAOR comprised many British units. Naturally, in Singapore they were ready to take on all comers in the open weight division. They never lost! In one final, they were up against massive Maori boys from the New Zealand Battalion. At first glance, there was no contest. An innocent bystander would have been forgiven for betting on the Maoris. They were just so much bigger and stronger. But, they were not smarter! Size and strength are not everything in a tug of war. The brute strength of the Maoris was no match for the superbly coached, disciplined and determined gunners.

To watch the British gunner tug of war team was to watch finely tuned machinery. All the men turned their bodies sideways, their faces away from the other team, their eyes and ears focussed only on their coach who stood to the side. He was their sole authority, the singular decision maker whose every command was sacred! He used his arms, held wide, to indicate to his team to raise, or lower their centre of gravity, and thus the rope. He was the only person to look at the opposition, who gauged their actions, their state of mind, their moments of weakness. He used a controlled, calm voice to order his team to dig in their heels and hold... and dig in and hold... and give one step and hold, take one step back... and hold... and another and hold, and save their strength, until that moment arrived that only he would sense, when the opposition might falter, and when that moment came, his voice gained a ruthless edge. Then, he would march them, stepping back, stepping back with precision, stamping their hob nailed boots, step after step after step, with an inevitability that crushed the opposition. Once marching back, nothing would stop them, and in a moment, those big Kiwi men would lie shattered, destroyed, humiliated by a bunch of tough, gritty, gunners. It was a battle of discipline, sheer will and finely tuned technique. It was a sight to be remembered.

Lo and I did lots of fun things in Singapore. Peter Kilpatrick, with whom I was in Vietnam, and his wife-to-be, Janine, and Morrie Evans, who was an officer in 106 Battery, and his pregnant wife Faye, and Lo and I all decided we would go on an island adventure. We had selected an island called Pulao Tiomin, which lies off the east coast of Malaysia. Tiomin was the island that was used in the filming of South Pacific. Deserted, wide, palm-fringed beaches with green-clad mountains behind, crystal streams and clear, clear oceans over stunning coral reefs... that is Tiomin—the closest thing to paradise I have ever seen. We set off from Mersing, a major port on the east coast, in a sampan that we hired on the spot. Loaded on board were our supplies for a week, including jerries of water, food, stretchers and a large Army tent. The sampan chugged away from the coast as we watched flying fish leap out of the waves alongside. It did occur to us that this was quite a long sea journey, so after the sampan had pushed itself up on the beach and we had unloaded everything, we took a great deal of time, using sign language, to ensure that the sampan captain would know when to come back and get us. His toothless grin assured us and off he put-putted. We were alone in our paradise. Life was glorious, peaceful and ever so romantic.

We fished and walked and just sat and chatted or floated about looking at the coral. One night us boys decided to paddle out to sea on the big inflatable raft that we brought along. Off we set, full of confidence. However, two things happened that caused us some alarm. Firstly, we noticed that the raft was actually going down. There must have been a leak somewhere under the water level. That's okay, there was a pump.

'One of us can pump slowly, the other two can fish... no worries,' said someone.

Then, to our horror, we realised that the fire that Lo and the girls had lit on the beach was fast getting smaller and smaller, not because it was going out, but because it was getting further away, fast! We were being dragged out by a strong current, heading who knows where?

'Right... paddle!' was the cry... 'And keep pumping. The bloody thing is still going down!'

We paddled and pumped our hearts out, for hours, and the light stayed about the same distance away. Eventually, it slowly started to become bigger again. At last it got bigger faster and we were released from the grip of that current. That was close. The girls were beside themselves, wondering where we'd got to. We could see their fire, but they could see nothing of us. They were right to worry. We had nearly become three men lost in a leaky raft, somewhere off the east coast of Malaysia. After five days, the sampan returned with its toothless captain, much to our relief. We were sad to leave that special place. Later, a tourist resort was built on that very beach!

Life was special over there. It was easy, we were spoiled. When it came time to go home to Australia, the tears flowed. We said goodbye to Ah Choo. We had grown very close to our amah and her family. We were invited to our amah's house for dinner, where we were served delicious food that we were quite sure they rarely ate, and where the whole family would stand and watch as we partook. What an honour it was to be their guest in their modest earthen floored house in the kampong. We had also invited Ah Choo's parents for dinner at our house, and on one Sunday we had taken them both on a trip across the causeway to the mountains of Malaysia, where they swam under a waterfall and picnicked amidst the Malay families. Ah Choo's mum had never been in a car before, and rarely been out of her kampong. She chattered in Chinese incessantly, continually pointing at things out the window. Now, at the airport, Ah Choo and her family, carrying large bunches of flowers, came to say their goodbyes to the 'Masta' and the 'Missee' that they had come to know and love as one of their own. They had taken us to their hearts just like we had taken them to ours.

It was December 1973, my two and half year posting was over. Farewell Singapore. It was an enriching and exciting chapter in our lives and in my Army career. We would never experience that lifestyle again.

13 LIFE AFTER SINGAPORE

'You know what? It won't stop raining and I reckon we're not going to get through,' I said to Lo.

We had already managed to thoroughly soak ourselves several times, but luckily the car was still dry inside. This was our second attempt to make it north after arriving home from Singapore. I was posted to 4th Field Regiment in Townsville, and, after collecting our Citroen GS—newly imported from Singapore—from the docks in Sydney, it was time to report.

It was the cyclone season, and it wouldn't stop raining. North from Rockhampton we became stranded on the highway between Ayr and Townsville. There was water from horizon to horizon as we drove through the cane fields with the sun going down. With the light fading, Lo walked in front of the car to spot any deep holes. She was soaked, miserable and scared. Then, to avoid the water rushing back up the exhaust pipe, I shut the engine down. Now the only noise was the sound of the two of us grunting and pushing the car through flood water that was flowing about a foot deep across the road, carrying with it the occasional tree. The only hope we saw was a slight rise in the road ahead, where we thought we might be able to position the car in an attempt to stay above the flood line.

Just as dusk was giving way to real darkness, we spotted a set of car lights coming towards us. Who could this be? Was there someone else as stupid as we were? Yes! There was! Here was a Land Rover coming the other way, with a family heading for Ayr, the place we had just left.

'Need a hand mate?' came the laconic Aussie understatement.

'Yeah mate that'd be good!' came the equally laconic, but secretly ecstatic, reply from me.

It was a close call. They towed us back to Ayr where we booked into what was just about the only hotel room left in town and stripped the inside of the vehicle. We had heard on the radio that the Burdekin River would flood that night, and it was the Burdekin that flowed straight through this town. So, what to do? Carry everything up to our room! Which we did: everything, including the car radio, and our new bed heads! These were special. Newly marrieds' bedheads! Can't get much more special than that! Well, we weren't that 'newly-wed' anymore, but these bedheads had rarely adorned our bed so far. They were made for us before we were married, but kept in storage whilst we were in Singapore. Up the stairs to our pub room they went, right past the public bar patrons who thought we were a couple of loonies. They were probably right. The river didn't flood.

The next day, all the stuff got carted back into the car, still under the gaze of the same beer drinkers who by this time were having breakfast at the bar, and we set off again. Avoiding the logs laying across the road, and the potholes, we limped into Townsville with soggy car, clothes and soggy just about everything that we had with us. Except the bedheads! They were high and dry on the roof! In Townsville, my nephew Rob Lans was relieved to see us. Rob, my brother's son, was our boarder for that year.

I was about to pay the penalty for spending three and a half years out of the first four years since graduation, on overseas postings.

I was introduced to my new Commanding Officer. Unimpressed—both me with him and him with me. He was not an inspiring man and the artillery regiment in Townsville seemed without purpose. I may have appeared to him as a cocky young officer. He was probably right. The CO seemed to dislike me. I couldn't put my finger on why. Possibly because I had been overseas so much. He hadn't even been to Vietnam, being of that age and seniority where he must have just missed the posting cycle, either as a Battery Commander or a CO. His dislike of me was confirmed by an incident that occurred within a few weeks of arrival. As a senior Lieutenant, due for promotion to Captain any time that year, I was tasked to take one of the batteries of the regiment, 108 Field Battery, to the High Range Training Area, just north west of the city, for some deployment exercises. Unbeknown to me, the CO had invited the Director of Artillery to visit that week and he took him to visit the 108 Battery gun line up on High Range. A gun line was a large and busy place with, in those days, six 105mm howitzers, piles of ammunition, six large trucks referred to as gun tractors, six ammunition trucks, plus command post vehicles, repair vehicles and all sorts of other support vehicles. All these were scattered around a piece of land about 100m by 150m.

In retrospect I'll admit, the gun position I had created was not a good look. It would have looked at home in the jungles of Vietnam or Malaysia, but not on the barren hills of High Range in far north Queensland. Here we were in Australia, training for conventional war, not the guerrilla war I was accustomed to. In conventional war, there are different imperatives that drive the make-up and design of a gun position—camouflage and dispersion. My gun position looked like it was at home in Asia, where it would have had an earthen bund around it, with machine gun posts.

The CO was furious, but he said little out in the bush. He decided to save up his anger and his punishment and waited until the battery came home that Friday afternoon. He had the Adjutant meet us at the gate. An Adjutant in a regiment is the senior Captain—the CO's Chief of Staff.

'Turn them around Mr Lans. The CO wants you to take the guns back to High Range and do the exercise all over again,' the Adjutant said.

'But Sir, I can see the wives and families lined up at the gate waiting to meet their husbands: they probably have plans, it's Friday afternoon,' I said.

I knew the Adjutant was only delivering the CO's message and I was sure that he was sympathetic to the situation, but he showed no sympathy.

'Turn them around, send them on their way, then go and see the CO!' was his reply.

Well, the CO gave me a solid earful, and sent me back to the bush to practise gun deployments, gun camouflage and gun drills until he was satisfied. It was another five days before we were allowed to come home. For a second time, we drove through the regimental gates.

The CO was angry with me and I was angry with myself that I had not seen this coming. I had been slack. I should have asked. It was clearly my responsibility. I guess the Senior NCOs in the battery didn't know me at all or they may have said that we should be doing this a bit differently, this is not the way we do things in Australia. Perhaps I displayed a cockiness that made them sit back and decide to let this bloke go and see what he's made of, let him get in the shit! That approach was not uncommon. It was the way that the NCOs and soldiers often tested their young officers to see if they have what it takes.

This was not a good introduction to the regiment. I called the battery on parade and I apologised to the soldiers for stuffing them around. They were not at fault. In the end, the fault was mine. As is so often the case, it is the soldiers who suffer when they are let down by their officers. But this was to be a good lesson for me and would help make a better officer out of me. It was the basis of learning to be a leader. All the while those same men in that battery, those soldiers and NCOs, watched and waited to see if I would stand up and become a leader.

I trained hard with that battery. We learned to deploy and camouflage like no other battery, and after six months we were doing well. Camouflaging a gun battery is hard work, all those guns and trucks all hidden underneath camouflage nets that are covered with layers of grass and tree branches with just a small hole for a barrel to peep out. It might take hours to perfect a position and just when we would be ready to have a rest and a brew, the order would come to move again. Even the camouflage nets were treated with a compound that prevented the infra-red signature of the guns and soldiers being revealed. The battery worked hard for me because we were becoming a team.

What I did manage that year, was my first adventure training exercise. Adventure training was in its infancy. Introduced to it by the British, who to this day remain the masters of adventure training, it was just beginning to be recognised as a genuine form of training in the Australian Army. I figured that a well-planned 'out of the ordinary' challenge would be a good distraction for the soldiers of the battery. Adventure training is about doing something a little different than the usual disciplined military training. It is where the guns and the ammunition and the command posts and the formal structures of rank are left behind, replaced by informality in unusual settings that present real, adventurous and sometimes life-threatening challenges. Adventurous training is designed to provide opportunities for young soldiers and young officers to stand up and take responsibility, for new bonds to form and, most importantly, to do so outside of the routine of the daily regimental slog.

In a giant leap of faith, I figured I could take the battery down a river somewhere and have an 'adventure!' The fact that I had only minimal river paddling experience was no barrier. The river I chose was the Tully River in North Queensland. I later learned that the Tully is one of Australia's foremost white-water rivers. I planned to paddle from Koombaloomba Dam up on the Atherton Tableland, above the famous Tully Falls, to where the river entered the sea. The small matter of getting around the falls that were over one thousand feet high, would be accomplished using helicopters to sling the canoes. The fact that neither I nor any of the helicopter pilots had ever slung canoes under a chopper before didn't seem to worry me. On top of everything else, I was also told that no-one had ever, according to the locals anyway, canoed the upper part from the dam to the falls, and if they had, no-one would have been mad enough to do it in open canadian canoes!

Undeterred, off we set, with canoes hired from Rob's school. We camped at the dam wall, amidst thick tropical rainforest. As was normal, we took along an Army medic and he gave us a few lessons in 'jungle' first aid. One of the very unpleasant plants that grows up there is colloquially known as the gympie bush. He had collected one from the rainforest for us to look at. It is an innocuous looking plant, about a metre and a half high on average, with large leaves which, upon closer inspection, have fine hairs all over them. Should you be so unfortunate as to brush against the leaves, these fine hairs will immediately adhere to your skin and inject a poison, which is intensely painful. The instant reaction is to scratch and tear at the stinging, which opens up the skin and allows even more of the poison to penetrate, making it worse.

'There are two things that you must do,' he said. 'Firstly, deprive the skin of oxygen by submerging it in water if possible, and, most importantly, look around the base of the plant to identify these smaller companion plants that usually grow where the gympie bush grows. The leaves from those plants, when rubbed into the skin, act as an antidote to the poison and will give quick relief.'

We absorbed the information, but nobody volunteered to be a guinea pig for a demonstration.

It was a memorable trip. Open canadian canoes are entirely unsuitable for a river which has masses of rapids and steep drops. The dense rainforest trees crowded the banks of the river in that stretch above the falls, making it difficult to slow down or stop to reconnoitre approaching rapids with any sense of calm or safety. Unable to get out of our canoes because of the overhanging vegetation, we found ourselves clinging to vines and tree roots just above the surging waters of every rapid, and there were many, so many that we managed to hole most of the canoes. We limped into our first camp above the falls where we spent most of the night patching holes. The next morning the Army helicopter successfully ferried the canoes over the falls. It was the first of many times that I would use helicopters on an Army Adventure Training Exercise.

There were other memorable times during that posting. Sometimes we would train by deploying along roads. There was no live firing. The aim of these exercises was to practice the movement and deployment drills and the associated logistics. On one such exercise, in which the Battery Commander had us driving through outback roads day and night, supposedly on some fast advance to contact the enemy, we were practising instant deployments, or 'crash actions' as we called them. In these situations, the battery might be driving along the road and would receive a call for fire from one of the FOs. On these dry firing exercises, these calls for fire were just for training and we would practice the drills, going through all the motions without actually firing anything. Whenever we received such a call for fire, the battery would screech to a halt, deploy the guns wherever it happened to be that we stopped, conduct rapid calculations to determine our location and the location of the target, and what settings to give the guns to ensure they hit the target, and then commence 'firing' the guns to support the action ahead. Once the fire mission was over, the battery would pack up and move on.

On one occasion, the battery was called into action at about 2.00 AM, on a remote public road west of Cairns somewhere, right next to a public rest area. In the rest area, there was a car with a caravan attached. Some unsuspecting tourists were calmly camped by the roadside, having no idea of what was about to happen to them. The call for action came! All hell broke loose! Vehicles screeched to a halt, gunners jumped of the backs of trucks, guns were detached and spun into action, NCOs started shouting, ammunition was offloaded, Land Rovers stopped all over the rest area and officers started yelling fire orders at the guns. This was quite normal for us. But then I wondered about the caravan and, sure enough, as I watched, the light in the caravan went on, then the door opened gingerly, and two elderly people slowly put their heads out of the door, then their bodies (still in their pyjamas) and they gaped: mouths wide open, dumbstruck! They must have thought that WWIII had broken out! Here, in the outback of Australia, in the middle of nowhere! I started over towards them to explain and apologise, but I saw that Vic Shields, one of my Bombardiers, had already reached them and was calming them down. Vic was one of the gentlemen of the battery, and he proved it once again as I saw him talk to these folks. Within minutes the crash-action was over, the guns were hooked up again with just as much noise, and off went the battery into the darkness. Those two people would have a story to tell their grandchildren, that's for sure!

Later that year I was promoted Captain by the Directorate of Artillery, and posted as instructor to Battle Wing, part of the Jungle Training Centre (JTC) at Canungra. I had not completed a full year with 4 Field, but was grateful for the opportunity to escape. It had been a learning experience.

Being part of Battle Wing in JTC, was such a change! It was a vibrant place staffed with young and enthusiastic Captain instructors and led by a gung-ho SAS Major. I remember my own indoctrination through the place just before my departure for Vietnam. Now here I was again, this time as an instructor. Very quickly I relaxed into a new role where my expertise was valued. This was my speciality as a young field officer—operations in close terrain. Although the war in Vietnam was over, the experience gained was not to be forgotten. Military units came from all over the world; from Canada, from the US, from the UK and from all over Australia, to be trained in techniques of jungle fighting and survival. Battle Wing comprised instructors from Infantry, Armour, Artillery and Engineers.

Some of my fellow instructors at JTC were to become friends for life, including Captain Peter Pursey, a 'Natio' who had decided to remain in the regular Army, even when National Service was abandoned by the Whitlam government. Peter and his wife Helen became our close friends. For a time, Lo and I lived on Mt Tambourine, which was a mountain village area near Canungra. Lo taught at Southport and Benowa Primary Schools. Each day she would travel that steep road, even when heavily pregnant with Matt, and each day I would ride my little 75cc postie scooter up and down the goat track which was the unsealed road that connects the Tambourine area with Canungra. I still marvel at the little scooter's ability to carry me up that steep hill. After a year, we built our first home at Daisy Hill, a suburb on the southern outskirts of Brisbane. It was here that Matt, our first son, was born.

On Battle Wing, I enjoyed instructing all soldiers, but particularly the US soldiers. It was an experience to supervise their squad runs in the morning, accompanied by that very distinctive chanting that the Americans do when running in a group. Supervising the squads on the obstacle course, clambering through pipes which disgorged them into mud holes or grappling with greasy cargo nets before climbing the tower at the river's edge for a jump into the icy Canungra River, was fun. The boss required us to do the obstacle course from time to time. When it rained and the troops got wet, so did we. When it got cold and the jungle greens that the soldiers wore proved completely inadequate, we as instructors suffered the cold alongside them, by not reaching for a dry jacket. When we ran them up the never-ending slopes of Heartbreak Hill, we ran alongside them. I vividly remembered being forced to carry that M60 Machine Gun up that hill on my pre-Vietnam course. It was no different now.

During my posting to Battle Wing, Matt was born. He was our first! We had only just moved into our new house at Daisy Hill and it was a bit of a battle for Lo, in a new house with no phone, no car (I had the car to get to work) and no family nearby. My family lived on the opposite side of Brisbane, unreachable for Lo during those little emergencies that happen with a first-born.

We were only in our house for a year, when, early in 1977, a posting to Geelong followed, as Adjutant/Training Officer of an Army Reserve Regiment. Geelong is a large industrial city an hour or so south west of Melbourne. This was unfamiliar territory for us, but Lo and I bundled Matt and took ourselves down there to establish a home at Point Lonsdale, a pretty, quiet seaside town not far from Geelong. The 10th Medium Regiment, a reserve regiment manned by part-time Army Reservists and trained by a small cadre staff of full-time officers like me, was led by an ex-regular officer called Tony Larnach-Jones.

Daily routine was all about preparing for the Tuesday night and weekend training sessions, and the annual 14-day camp at Puckapunyal. The regiment was housed in very old barracks that were situated next to the Geelong Prison, sharing a boundary wall with the prison, which was in fact the prison's outer wall and was patrolled by armed guards who watched from the towers at the corners. The cadre staff played a lot of sport and we entered a team in the local Geelong volleyball competition. Some of the Army reserve members were guards at the prison and they told us the prisoners also had a team, naturally not in the local competition. They were desperate to play a team from the outside, so we volunteered. The only proviso was that they would not play 'away'—we had to go there. That was understandable! We were invited to enter the prison yard and play against them 'on the inside'. That had its own challenges. Somewhat tentatively, we were led through the massive, century-and-a-half old gates in the thick bluestone walls into this prison that was one of the oldest still in use in Australia. The prisoners were on yard release, so the cells were empty.

The sight of the prison cells silenced us all. The thought of men living their day-to-day lives amongst these cold, dark walls in small cells that permanently smelled of urine was sobering. Even in the 1970s, plumbed toilets had not been installed into these buildings with their foot-thick walls, so the prisoners were still using personal pots in the corner covered with a small cloth. It was eye opening. When we played the game in a large walled yard that was full with at least a hundred prisoners all wildly cheering for the home side, we felt small and vulnerable. This yard also housed showers; open air showers, in Victoria! As we filed past there they were: dozens of blokes having showers, not more than 30 metres from the volleyball court. Some of the prisoners chatted away about all sorts of subjects, but mostly about how long they had to go until they would be allowed out.

The game was memorable. The referee was one of the prisoners, and the linesmen were the prisoners watching, all of them! When a ball was in question, dozens of voices would call to advise if it was in or out. Mostly it was up to the side where the ball hit the ground to decide whether it was a good or bad ball. But on one such call, there was a bit of argy-bargy going on, some friendly argument over whether or not a particular ball had been in or out, when one of the fellows on our side said loudly:

'Well, we're honest... don't know about you guys!'

Stunned silence! Everyone drew their breath! Every prisoner eyeballed us, ot a word was said. The guards on the tower even lifted their rifles into a ready position. We stared, someone do something!

'Ha!' said the offending player on our side. 'That got you, didn't it? Your point of course. We knew that!'

Wise decision! With that, the tension was broken and the game resumed.

As it turned out, one of the soldiers in the regiment would soon be joining the prisoners next door. Three of us cadre staff members lived on the Bellarine Peninsula from where we drove in to work together. As we entered the barracks one day, we noted that there was a large hole in the outer netting security fence, and that the wall of the Quartermaster's compound had been smashed. Then we realised that we could look straight in to the armoury where the regimental weapons were supposed to be. They were gone! We've been robbed! The police were called:

'Jeez mate, all those weapons stolen from the Army? Someone is going to be cranky about that,' said the cop.

Someone with a knowledge of the armoury and the layout of the regimental stores, had smashed their way in through the old stones of the establishment walls and simply destroyed the armoury with sledge hammers. Obviously, the mortar that held the bluestone together no longer worked. They made off with a variety of 7.62 mm SLRs, some light sub-machine guns and other bits and pieces. Some thirty or so weapons in all. Who knows what that they might have fetched on the black market! The armoury held no ammunition, luckily.

We were a bit shocked, wondering where all this would lead when, within a very short time, the police rang. Would we please come and see them? There was someone they wanted us to identify. Incredibly, when we arrived at the police station, there sat one of the young gunners of the regiment, looking very sheepish, with two other blokes alongside that we did not know. As it turned out, the police had very little trouble catching them. In disbelief, we listened to the story. They had hired a van, using a *real* licence and a *current* credit card, loaded the van up with the weapons that they stole, then unloaded the weapons at the *regular* address of the license holder. Then, if that wasn't enough, instead of returning the van to the rental place, they decided to *dump* it in Corio Bay, which is the arm of Port Phillip Bay upon which Geelong is located. Why they thought this necessary, you would have to wonder. Perhaps they had seen too many movies. But, if they had, they'd obviously missed the movies where the vehicle doesn't sink, because in this movie re-enactment, the van *floated!*

It was quickly spotted and reported. Police were called, and when the van was opened, they found a rifle that had been left behind! Yet another mistake in this finely planned heist! All this within minutes of us ringing up and telling them about the break in. The police simply linked the rental van, the license, the credit card and the stolen weapon, and in no time, they were outside the real residence where the rest of the weapons were hidden. I remember looking into the eyes of the young gunner and he stared at me with a look of incredulity, as if it was so amazing that he had been caught. Clearly not very smart.

He and his mates were off to the prison right next door. Luckily, they didn't make the prison volleyball team! As a result of the break in, we lost our accreditation to store weapons. The Army quickly changed its policy in relation to the conditions, construction and safety of armouries. No longer was it acceptable to store weapons in a 150-year-old facility.

It was at Point Lonsdale that Raewyn, our second child was born. She was a teeny little thing and came very quickly, so fast, in fact, that I hardly had time to get Lo to the hospital before she was born! Then towards the end of my two years there, in 1979, I was posted once again to 4th Field Regiment in Townsville. It seemed our family experiences with the tropics and my experiences with that particular regiment were not yet over. My reception in Townsville put paid to any doubts I had. I felt welcomed back. All trace of those earlier days of uncertainty had gone, those days of transition when the Army was struggling with the loss of the Natios and the loss of identity after the cessation of the Vietnam war. Gone were the officer personalities that had prevailed in 4 Field Regiment previously. I recalled the officers that served in the regiment at that time and, on reflection, none, or perhaps only a few of them, seemed to be about in the Army at all. I could not detect any low morale. There was a lively new vibe about the place. I was welcomed as a long-lost son by the CO, Lieutenant Colonel David Bedford, who was the ex-Battery Commander of 106 Field Battery from my Vietnam days. I learned that he had personally asked the Artillery Directorate I be posted to his unit. How amazing! He actually remembered me from when I thought I was a young and insignificant 2nd Lieutenant. It was evident that my arrival had been planned for some time, because a succession plan had already been put into place: I was to be posted as Battery Captain of 108 Field Battery until the end of that year, when I would be promoted to Major, and take over as the Battery Commander. This was exciting stuff! At last a chance to again serve close to soldiers and learn more about command and leadership.

Alan McClelland was the incumbent Battery Commander. I walked into his office and he met me with an intense gaze. It was disconcerting. He was a man who knew exactly what he wanted, both from his battery and from his officers. I was a bit hesitant with Alan at first.

'This battery will be the best battery in Australia,' he said to me, 'and we will prove it by winning the Divisional Artillery Commander's award for best battery. I expect you to see to it and carry it on when you take over.'

No pressure! He told me that every year, all the batteries in the Army competed for the award by completing a series of tests, designed by the Divisional Artillery staff, who would fly all over Australia to conduct the tests. Before I arrived, 108 Battery had done well, but they had not been declared the best. So that was yet to come. 107 Battery, the other gun battery in the regiment, had won that year. Okay, I thought, so I have something to aim for here. I felt energised and inspired and I walked onto the BC's balcony and looked out over the regimental area and across to the CO's office, and thought about that time a few years ago when I, as an inexperienced young officer, straight out of cadet school, intimidated by the sheer scale of the unknown that lay ahead of me, had started my association with this regiment, in Vietnam, and in Singapore, and then a few years later here in Townsville again.

I walked to the gun compound, where I remembered all those families looking at me and blaming me for sending their hubbies back onto High Range. But this was a new start. I had a lot more experience. The regiment had an inspiring and competent CO and the post-Vietnam hangover was over. I wandered about the barracks, meeting the soldiers, sizing them up as they sized me up. I remembered many of them and I felt welcomed. It was a promising start.

This was going to be an exciting time.

14 108 FIELD BATTERY

We happily settled into our married quarters in Tam O'Shanter Drive, Kirwan. The family now comprised Matt and Rae and our German Sheppard dog called Jo. We thought it was the ant's pants, that house! Elevated on stumps to let the cool air flow underneath, as was the way of houses built in the tropics in those days without air conditioning, it was the cream of the Army's new housing stock. The house was very near to the grasses and shrubbery of Townsville's outer surrounds, that were full of brown snakes. It was here that Robbie, our third child was born. Once again there was considerable pressure placed on Lo with a new little boy to add to the two young kids and the dog!

4th Field Regiment Officer's Mess in 1981.
Photo Australian Army

For me, the posting to Townsville was a highlight of my career. This was what I reckoned being an officer was all about: leading men! I was in my element! The down side for the family was that I was out on exercise for more than half the year—every year. So, for Lo, who was bringing up two, and later three little children in a town thousands of kilometres away from her sisters and father and mother and my family, this was a difficult and challenging period in her life. I was often out on Army exercises with 'the boys', my men, doing my job, having a great time. She was mostly home with the kids, changing shitty nappies, without even a telephone to call her mum. Townsville was still a growing town then. We loved it though. The climate was hot and humid for much of the year. But we had the mountains and rivers to the north, where such mighty rivers as the Herbert, the Tully and the Barron flowed, and we had Magnetic Island where, for just a few dollars, Army families could hire cottages and swim in the crystal waters of the box jellyfish-free Horseshoe Bay. As was the way of the married quarter patches of the day, all around us were other military families, so at least there was some support for Lo when I was away.

Major Al McClelland, having heard that I was a keen on being an adventure trainer and river 'expert', decided that I should put together an adventure for the battery. We decided to canoe the Herbert River. Al knew nothing of the Herbert, didn't even know if it had been done before, but thought it a good idea in that typical Army manner of taking on anything.

'... and besides, Ben, you're a bit of a canoe expert, aren't you?' he said.

Well, not really! My expertise came entirely from those early days in 1974, with Lo and my nephew Robert, when I learned a little about the dangers of rapids, the stoppers at the bottom of the water falls and drops, and the way to pick a line through fast flowing water. My Tully River experience from a few years ago helped, though I was far from being an expert. But, I wasn't going to miss an opportunity like that!

The adventure was on! There we stood, a bunch of soldiers from 108 Field Battery stationed at Lavarack Barracks in Townsville. We were artillerymen:

gunners, about to canoe the Herbert River from the crossing to Blencoe Junction, about 100 kilometres downstream, past the infamous Herbert River Falls, through the gorge that followed, and past the many rapids and falls yet to be identified. We knew there was a lot of white water, but we knew little of optimum river conditions or safe water heights. The local land owners had told us that no kayakers had ever done Cashmere to Blencoe in one go before. They said some sections had been explored using whitewater kayaks. But, were going to do it in open canoes! These blokes were good at manhandling 105mm howitzers, but not canoes. This major consideration aside, we all looked forward to some hard work and unknown challenges, the camaraderie of campfires and the odd rum toddy. This was what adventurous training was about wasn't it? What could go wrong? They were agile and strong; I reckoned they'd learn quickly. The exercise group consisted of the river party, comprising nine open canoes plus one whitewater kayak paddled by Al, and a vehicle-based communications party, which would roam the tracks in the vicinity of the river as safety back-up.

Although the river was only a few metres wide at Cashmere, upon entering the water most of the blokes managed to point their canoes in every direction except downstream. A bystander would have been excused for asking which way were we heading. But, after some splashing about we set off, zig-zagging down the river. The first few days were glorious. Blue skies, sandy banks, shady gums and gentle races. Then the rapids started. Al showed the way by completing the first one upside down. We all cheered and whooped, but most of the blokes didn't do much better. Invariably at the end of each rapid, the sight of drums, paddles and assorted camping gear preceded an upside down canoe empty of its occupants, who would appear soon after, bedraggled and laughing. Should I have been worried?

Knowing that the Herbert River Falls was a 'show stopping' obstacle, I had arranged to get help from the Townsville-based 162 Helicopter Reconnaissance Squadron. A chopper would lift our canoes over the falls.

On the third morning, an Army Kiowa Helicopter landed above the fall. The plan was to sling the canoes one by one down the face of the fall onto a small beach beside the pool at the bottom. First the pilot dropped one of the gunners at the bottom of the fall to unhook each canoe as it was delivered. Then, the rest of us watched warily as one by one the canoes gyrated in the spiralling air currents of the deep, narrow ravine.

Canoes and equipment safely in place, the pilot ferried the rest of us down to where we were soon settled around a huge fire. Irish, one of the Bombardiers, caught our first eel and Vic, our radio communicator, rubbed handfuls of salt into it, wrapped it in alfoil and threw it in the fire. Perhaps it was because the eel came from a crystal-clear river or that we had ravenous appetites, but it tasted like chicken. What a meal! We slept soundly that night, our bodies re-charging for another great day.

Army Kiowa Helicopter lowering the canoes around Herbert Falls.
Photo Author

'Hey boss you awake?' A voice came from somewhere. 'Nah', I said jokingly. 'Let me sleep. Go away!'

'Did you see that big waterfall around the corner?' came the voice again.

'Of course I did, you great pillock, we just flew over it yesterday!' I said.

But the voice came yet again. 'Not that one, there's another one just around the corner!'

'Nah,' I said. 'We recce'd this river from the air and there's only one big fall, the one that's behind us.'

'Horseshit boss!' Came the voice. Now I was awake!

I had done a helicopter recce a month or so before, accompanied by Ray and Greg, two of the young officers in the battery, but we hadn't spotted this second set of falls.

During the recce, when we were near the Herbert Falls, the pilot had spotted a huge crocodile in the river, and had thrown the chopper into a seriously steep dive to show us. Unfortunately Ray, who suffered from air sickness, responded by vomiting into his hat! Mayhem! None of us got to see the crocodile. In fact we were so distracted by the whole manoeuvre and the smell emanating from Ray's hatful of vomit that we momentarily stopped looking at the river at all and didn't even see a second set of waterfalls. Anyway, that's what I claimed later.

Sheer walls and steep drops. Which way? Luckily Barry, one of the gunners, had risen early and had already recce'd a potential route using a small platform to the right of the main fall. I took a look. 'Okay fellas, Barry's plan is good. Each canoe will deliberately enter the fall's pull then turn right and paddle hard towards the platform. Approach too fast and you'll charge over the edge, too slow and you'll be sucked over. We'll secure a safety rope which will be flung to each lead paddler, but not until you have turned. You will have just one chance to catch it!' I said, heart thumping somewhere around my Adam's apple. 'Room is limited so when the first canoe is secured we will immediately start lowering them down next to the fall.'

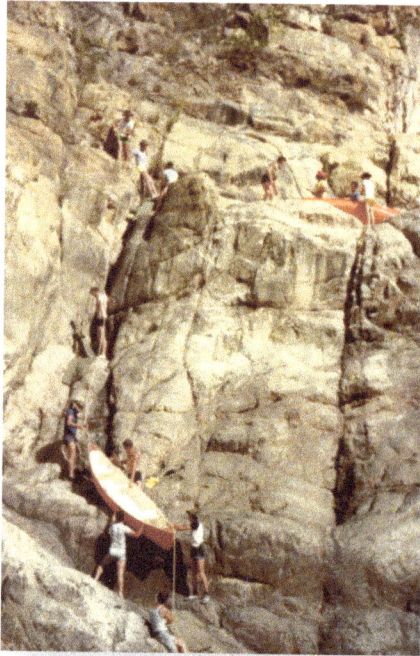

Passing the canoes down part of the 50 metre drop to avoid the second set of waterfalls.
Photo 108 Field Battery Archives

It seemed to take hours as each canoe rounded the bend and secured the landing, followed by the slow and deliberate process of roping the canoe and contents down the vertical 40 metre escarpment in stages, with some of us perched precariously on the rock face to lift loads over outcrops. Many expletives and crushed fingers later we finally got everything down. It had taken almost all day. We made camp around yet another campfire. The blokes immediately began kidding me with... 'Right, well where is the next waterfall then, boss?' That was the catch cry for the rest of the trip.

The next morning we became aware of a change in river character. Gone were the scratchy bushes. Instead we dealt with sand, rocks and sun-bleached pyramids of large car-sized boulders. Scattered amongst these pyramids were piles of round stones, ready to break an ankle with one wrong step. We were now in a gorge where the sides were at least one hundred metres high. Gone was the cooling breeze. The sun was threatening to peep over the escarpment at any moment and once it did, we would bake! And the river was no longer babbling over shallow races; instead, it had become a series of crystal pools lined with sand and pebbles; tiny, shiny spheres that reflected the sun. The pools were generally between 50 and 100 metres long with no apparent surface flow, ending in jumbles of smooth river-rocks, which signalled a drop in the gorge floor to the next level and the next pool. At the end of each pool, the water seemed to disappear by ducking under and around the rocks, re-emerging in the next pool in many small trickles or large swirls. Sometimes the water converged sufficiently to create pressurised jets of water that squirted their way over the rocky drops and sometimes the water flow even became sufficiently strong to create a small waterfall of a metre or so, or a small rapid. In this way the river slid, surged and splashed its way downstream; bubbling, singing and swirling along. This did not make it easy to get downstream. Portaging our canoes and our kit over, around and sometimes through the piles of boulders became the standard, exhausting dawn to dusk routine. Unload, carry the gear, drag the canoes; then repack, paddle a hundred metres or so and... unload again, on and on. All the time the sun baked us red. We were thankful for the cool water.

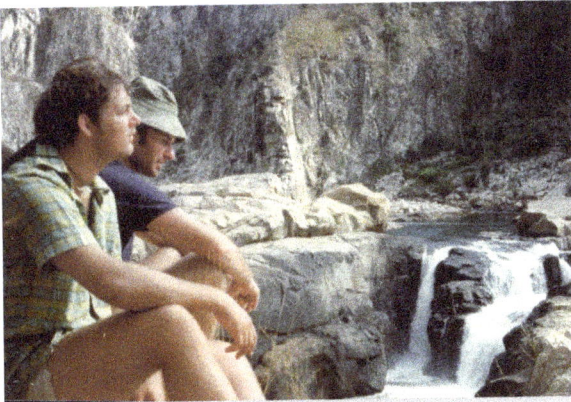

Captain Ray Cook and Major Alan McClelland overlooking the 50 metre waterfall that was (improbably) missed during the helicopter reconnaissance of the river. The canoes were portaged along the shelf to the left of the smaller shute of water and lowered down using ropes.
Photo Author

After the extremes of portaging and pitting ourselves against a river not wanting to be paddled, I began to worry that fatigue might cause us to make a bad decision. Three days after the falls we had to lower our canoes around, with at least another three days to go, we came upon a rock face against which the water seemed to come to a stop. Powerful welts of agitated water surfaced only to be sucked down as the next welt boiled to the surface. Angry water surged around the rock face to the left but I couldn't see what was around the corner. I hesitated: stared at the bubbling surface. The prospect of portaging our canoes around using the sides of the canyon was daunting, so paddling around was the easy option. Although my gut said otherwise, I decided we'd try it.

A couple of the boys held us back while Webby, my paddling partner and I sat, for what seemed like an eternity, as the canoe jumped about crazily in the savage water. I pictured us being tipped out and sucked down, tumbling over and over, held in powerful boils deep below, and suddenly I saw reason.

'No go!' I said. 'Too risky!'

I could see Webby's shoulders slump with relief.

That night I slept the sleep of the exhausted, waking only to throw Al's river shoes into the fire in the middle of the cold night, mistaking them for firewood. Melted rubber soles in the dying morning ashes were the only testimony that his shoes had ever existed. He was furious!

'Thank you for sacrificing your feet to keep us warm, Al!' I said cheekily, knowing full well he was about to demand my shoes as payment. Luckily one of the gunnies had a spare pair and so I was spared the indignity trying to haul my gear over sharp, baking hot rocks, without any foot protection.

Towards the end of the exercise, a memorable event happened. A British officer, who was out from the UK on a three-month exchange and who was reputed to be a bit of an adventure training specialist, dropped in on us. He was partaking in the annual 'Exercise Long Look', which was a formal exchange of officers and NCOs between the British and Australian Armies. Al and I knew he was coming and for whatever reason, we had been told he was an outdoors expert, specialising in river adventuring and canoeing.

We left instructions for him to join us in the field as soon as he arrived, and he was literally 'dropped' on us by helicopter, having only just arrived from the UK. Captain John Brown was his name. He was a British Army Artillery Officer. There he stood in his khaki buttoned shirt, peak cap and pressed trousers, looking bemused in only the way the British can. The fact that he had travelled halfway around the world, to be dropped into a river in the middle of nowhere did not perturb him too much.

'All in a day's work for an Army man,' was the first thing he said.

'G'day mate,' Al said. 'Welcome to 108 Battery. Here, get that uniform off, hop into some shorts, send your bags back with the chopper and hop into a canoe. Have you ever done this sort of thing before?'

'No!' was the short answer.

'What?' Al said. We were shocked! 'But you were supposed to be an expert!'

He wasn't, but like the good Army officer he was, he did as he was told and just got on with it. What was probably the greater shock was his introduction to Australian soldiers. The informality of us Aussie soldiers, particularly when in the bush, has been, and is, legendary. Our soldiers are often irreverent, less inclined to accept authority without question, and intolerant of incompetence. They will exploit any perceived weakness in their leaders by simply challenging them. But they do not lack discipline. They will work hard to complete assigned tasks no matter what, provided they understand why and have confidence in their leaders. I think John Brown found the gunners of the battery a bit confronting and in his face. But he was a likable bloke. He whipped off his shirt, got stuck into the daily grind, and in no time, he looked like an Aussie bushman, burnt to a golden brown by the savage Queensland sun and the high humidity.

The exercise ended after ten days with a huge climb out of the gorge at Blencoe Junction. We dragged the canoes up the forty-five-degree walls of the canyons for well over 200 metres in height, to reach the truck rendezvous point, being the first and probably the last to do the Herbert from Cashmere to Blencoe in open canoes!

We were bruised, sunburnt, hungry and exhausted, but very pleased with ourselves. I was proud of the gunners for what they achieved. Not one grumble. This challenge had moulded young and old; experienced and inexperienced individuals into a team that worked together and relied on each other. The river was capable of breaking anyone's spirit, but their belief in themselves never faltered. This was an important step towards facing the unknown challenges and potential adversity of life in the Army. The stories from this trip were sure to be told again and again.

However, that story doesn't end there. A couple of years later, when I was on the staff of the Ministry of Defence in London, sitting in a large audience of intelligence officers waiting for a presentation to start, I heard a voice behind me. Someone happened to be talking to a person sitting next to him, about an incident that happened when he was on exercise 'Long Look' a few years ago, where he had been dropped into the Australian outback to join an Australian gun battery. This battery was at that time in the middle of an adventure training exercise, paddling some river in Queensland. I was listening to the conversation because it mentioned places familiar to an Aussie. I turned around and could not believe it. There he was, (now) Major John Brown! When he realised who I was, he nearly fell of his seat! An unbelievable coincidence! We fell into a back-slapping frenzy of storytelling, much to the annoyance of the presenters, who were trying to get on with the day's work. As the only Aussie on the staff of that headquarters in London, I was known to be a bit of a larrikin. Actually, every Aussie on exchange was expected to be a joker and a stirrer, and this loud raucous laughter, was another example that the Aussies were just a bit uncouth! I am not sure if the fellow to whom John Brown was talking actually believed all this or thought it was a set up.

At the end of that year, 1979, I took over from Al and was promoted to Major in command of 108 Battery. I did so with a running start. I already knew the men and they knew me. I felt this was the beginning of what was to be a very rewarding experience. I could feel it in my bones.

15 LEADING AND LEARNING

Being the leader, being in command, is something of a frightening concept for many people, but for us in the Army, it was our bread and butter. For those of us in any of the services, having a commander to answer to and being in command is what we do. Soldiers are promoted to become junior NCOs and then senior NCOs as they work their way up. Some of them are given a commission and go on in officer ranks, but most officers enter via the officer training schools. Unless you are the Chief of the Defence Force, there is always a commander above you, and let's face it, even the Chief of the Defence Force answers to the Minister.

As you work your way up the ranks you are expected to execute command, over more and more people. One of the strengths of the Army, and the services as a whole, is that it develops young people who are not afraid to lead, make decisions, command, accept and discharge responsibility. This responsibility of command however has many facets. Command implies leadership, guidance and knowledge of who and what you are commanding. The system within the Australian Army works well. Those who are in command have earned their place, not assumed it some through class, birthright, corruption or another form of reward. Australian soldiers, in my opinion, respond well to being under command and taking command. The character of the Australian soldier is unique and perhaps it is so because of the history of this country, the stock that made Australian soldiers the resourceful, no-fuss and up-front individuals they mostly are.

Generally, and I speak of today's soldiers, the average Aussie Digger thinks for himself, is keen to do well, and does not suffer fools gladly. The young men and women that make up our voluntary forces today, appear to me to be of the highest standard. I have gained this impression not because the news tells me so, but because I have been lucky enough to continue to witness many soldiers and other servicemen and women at their work, and I remained closely involved with the Defence force throughout my working life. As I witness the young soldiers of today going about their tasks, I see the same comradery, the same pride, the same spirit that was about when I was a young officer fifty years ago.

At the end of 1979, about the time that I took over the battery, Lieutenant Colonel Bedford was replaced by Lieutenant Colonel Ian McGuinness. I didn't know him, but he seemed a good man. He turned out to be a true gentleman. His philosophy on commanding the regiment was to let his Battery Commanders run their own show, as long as it achieved the overall outcome required by him. He would allow us to manage our own affairs without interference. Excellent! The other gun Battery Commander, commanding 107 Battery was Major Peter Kilpatrick, my friend with whom I had shared various adventures in Vietnam and Singapore. As Battery Commanders, we respected the CO's wishes, appreciated his calm hand at the tiller and learned from his measured response to every drama, big or small. I consider myself fortunate to have served under such a CO. Second-in-Command was Major Ron Lenard, a nice affable officer, who cared a great deal about the officers. He was unlikely to interfere. Even better! So, we were in good hands. I was lucky. I didn't have to serve with any of the 'gunner zealots' that were about, and there were quite a few of them. These gunner zealots I describe as over-ambitious officers who cared more about their careers and the outward appearance of the regiment than the well-being of their soldiers.

Having good and loyal Senior NCOs is a vital element of a good unit, as is having an understanding and supportive CO. I was lucky on both counts. Having good officers is also important, but most young officers can be moulded and shaped into good leaders and if they can't, then they need to be removed and taken away from leadership roles. Not all officers have to be leaders. Some can become administrators, logisticians or used in positions where they have little command responsibility.

The post-Vietnam period had been something of a let-down for the Army. No longer was there the political imperative to have a hardened force ready to deploy, where units cycled through a year of preparation for operational deployment, completed the deployment, then came home to recuperate and get ready for the build-up again. There was the Cold War going on between the west and the east, but that didn't really seem to impact on us, and the funding that was allocated to the forces seemed to shrink with every budget. Equipment was allowed to run down. Personnel numbers were allowed to run down. Recruiting was at a low level. Sometime in the late 70s, the politicians realised that the Army was in such a state that, if there was a sudden need, there was actually no part of the force that was fully trained, equipped and manned, ready to go. To remedy this, the government decided that there would be such a thing as an 'Operational Deployment Force' or ODF. We in Townsville, as part of the 3rd Brigade, would be the ODF. One infantry battalion and one gun battery plus some engineers and some logistics people and units, would be required to be on standby for overseas deployment with varying degrees of short notice. This was good, as it meant we had a priority on things like ammunition and equipment, vehicles and also recruits that were marching out of the Kapooka Recruit Training Battalion, to be allocated to units.

Being in a fighting arm like an artillery regiment is like being in a club, an exclusive club. Members tend to stay loyal to their regiments. The greater the comradery, the greater the spirit, the culture and the confidence of that regiment, the more loyalty is generated. This is fed by the professionalism,

Part of the Battery Commander's Party, my support team when on exercise in the field, having a quick meal under a camouflage net. From the left Gunner Ian Biggs (driver), Gunner Tony White (signaller) Sergeant Harry Pregnell (Battery Commander's Assistant) and Bombardier Spaulding (Command Post Operator).
Photo Author

the standard of training, the achievements in the field and the care that is shown by the officers and NCOs for their soldiers. I have always believed that if a regiment takes itself too seriously, things are likely to go wrong. Training to a fine degree of competence and readiness, whilst maintaining a sense of realism in a Defence Force that actually has no operational deployments to go on in the foreseeable future, is a challenge. This challenge is not always well met by commanders. When training becomes a slog, a bore, soldiers get fed up and leave, or play up when they are not supposed to. Commanders, just like their soldiers, need to understand that when it is time to play, soldiers play! But when it is time to work, soldiers work! There must be give and take. Our CO, Colonel McGuiness understood that, and allowed each of us Battery Commanders to devise our own programmes to ensure that the soldiers were ready, well trained and most of all... happy.

One of the perks of being in command is that I managed to get many, many helicopter rides, all over the countryside and for lots of different reasons, including reconnaissance, adjusting artillery fire or simply just to get from one place to another quickly. One day I was in one of the newish Army Kiowas, which is just a military version of a Bell Jet Ranger. I was the only passenger and was sitting in the front seat, chatting with the pilot about all and nothing on our way to a reconnaissance somewhere up north. Suddenly, the peace was shattered as one of the helicopter alarms went off with an ear piercing, all penetrating noise that forced us to pay attention instantly. I looked at the pilot and immediately realised that we were in trouble.

'Hang on' he yelled!

And that's all he had time to say as he immediately went into his emergency routine. We were flying at about 1500 ft, which suddenly seemed pretty high to me, but he cut the engine and immediately threw the helicopter into a steep downward spiral, using the remaining speed in the rotor blades as a braking force to prevent the aircraft from plummeting straight down. Now 1500 ft seemed even higher! As we approached the ground he once again yelled:

'Sit tight! This is going to be rough!'

It sure was! Wham! We smacked against the ground with a force strong enough to bend the skids. Blades were still turning. The pilot threw switches everywhere! And then, we just sat, white faced, until the blades stopped. I looked across at the pilot and his eyes were wild, fixed into an unseeing stare by the adrenalin that was still pumping around his body.

I was silent, still shaking. Or, more accurately, I *started* shaking now, having had no time beforehand to actually register what may be happening, my body now decided to go into mild shock!

At last the pilot started talking:

'Shit, mate... that was transmission failure warning! The magnetic filter must've detected a metal fragment, and mate, that's a sign the gearbox is about to implode! Next thing would have been jammed blades and us... cactus! I would've had bugger all chance to do the auto rotation under control. We were bloody lucky I tell you!'

So, there we were. Somehow the pilot had managed to send out an alert and in an hour or so, another aircraft came from Townsville to ferry us back.

'Well done, mate!' I said. 'You saved our lives!' What an understatement! No recce that day!

One of the fun exercises that the regiment took part in was an annual command post exercise held in, of all places... Hawaii! For this, the headquarters of the units that comprise the 3rd Brigade in Townsville, went to Schofield Barracks on Oahu Island, Hawaii, and joined up with the units of US Pacific Command (USPACOM), stationed in Hawaii. A command post exercise is one where only the officers and their command post operators and radio signallers take part. The aim is to train the commanders and their staffs in making decisions based on a scenario that is simulated, and the actions that require decisions are stimulated by control staff—controlling organisations that represent allied or enemy formations. These exercises were the forerunners of the electronic, computer-driven live, virtual and constructive simulation systems used in training today.

We went to Hawaii and had a ball over there. With Hugh Polson, my friend from Portsea who was also posted to 4 Field, and a few other officers from the regiment, we hired a huge 'yank tank'—a bright red convertible Cadillac. What a car! We put little Australian flags all over it and drove around the island as if we owned it with our flags proudly fluttering! We made short work of the Marines who were on guard at the gate post of Schofield Barracks. They were so confused by these larrikin Aussies in their flashy car, that they would salute us anyway, just in case we happened to be important! The actual exercise was only a few days long, but we were in Hawaii for three weeks. At one point, after a particular heavy night, I lay down on the floor of the command post in a quiet moment, just to catch up on a few 'zeds'. To my horror, I woke up to the sound of my CO's voice, talking to a person who was clearly a very important person, judging by the sound of deference in the CO's voice.

I peeked out from under my hat and saw two pairs of boots inches from my face. The two of them were standing there chattering and completely ignoring this body on the ground at their feet! Bugger, what to do? Luckily, we were operating from a tent which had low tent flaps near the ground, and on this tent the flaps had not been done up. I rolled over and out of tent, out of their view, got up, brushed the grass off, and stepped into the entrance as if I had just arrived from somewhere. Without missing a beat, the CO introduced me to the visiting US general, who looked me straight in the eye and asked me how did this Aussie command post differ from the US ones? As I gathered my thoughts and started gushing forth with some considered comments, I wondered if he knew it was me lying at his feet a few moments ago. All the time I had to resist looking at my soldiers, who were out of direct line of sight of the general, and who were having incredible trouble not guffawing out loud and doubling over with laughter. I still don't know how the CO managed to keep his face so straight! He never said a word later. An extraordinary man was my CO!

I was very proud of the spirit and the professionalism of the officers I had around me in the battery. They were a good bunch. There was always the ability to manipulate getting a good officer into the battery when you spotted one. The good ones stand out because it is nearly always in their attitude towards soldiers that makes or breaks an officer. Sometimes the CO would allocate a new officer to the battery, with the warning that this one was in need of 'sorting out'. I loved that challenge! Generally, it would work out and the wayward young officer could be shown the 'errors of his ways'—a favourite saying in the Army. I had been shown the error of my own ways a few times as a young officer, let's face it!

On one occasion, I had taken my officers away to support an activity somewhere in Sydney and we managed to get ourselves stranded at RAAF Richmond, the Air Force base just west of Sydney. I had Cookie and Stevo with me. Stevo was Captain Ray Stephens, another of my FOs, and he and Cookie were trying desperately to organise a RAAF flight back to Townsville for the three of us. Civil flights were pretty hard to justify those days. The RAAF flew regular Hercules aircraft flights all over Australia and servicemen were simply expected to use them, not use taxpayer's money to buy airline tickets, as happens these days. However, in those days, making a 'long distance phone call', or 'STD' call as it was known, was expensive. In the services, unless your posting was such that you were specifically authorised, you had to spin a story and request the telephone receptionist to connect you. Cookie and Stevo were in the telephone box and I was standing outside it. Cookie considered himself a bit of a ladies man. He was kind of cool, or so he thought. Stevo on the other hand was a joker. He was an absolute comedian who never took anything seriously and had a smart remark about everything in life.

Cookie, whose real rank and name was Captain Ray Cook, was on the phone, and I could hear him trying to persuade the receptionist to let him make a long-distance phone call to Townsville, pleading with her. I saw him hand the phone to Stevo, then it went back again. Then, suddenly the phone was slammed down. Pandemonium broke out! Cookie had taken his bush hat off and, in the confines of the telephone box, was slamming it across Stevo's head. Stevo promptly fell out of the phone box onto the ground, doubled over with laughter.

'What happened?' I asked.

'Sir,' Cookie said to me, pointing to Stevo, 'That bloody idiot just ruined our chances to get a lift home on a Hercules aircraft tonight. The telephonist asked me my name, but she laughed when I told her, so I was trying to explain to her that Captain Cook was my real name! Well, I said that I had a fellow officer here who'd testify to the fact that I really am 'Captain Cook!' So, I handed the phone over to Stevo, and what does he say? He says: 'Hello sweetheart, it's Napoleon here... what seems to be the problem? Bloody hell! She hung up on us!'

By now I was also doubled over with laughter. That was classic Stevo. The saddest thing is that in less than a year, Stevo's life was to change forever...

With Captain Craig Stevens and Captain Ray Cook at Richmond Airbase discussing options on hitching a ride to Townsville on a RAAF Hercules aircraft.
Photo Reginald Lans

I am sitting in a Huey helicopter next to the CO of 2ⁿᵈ/4ᵗʰ Battalion. I'm his Battery Commander and coordinating the fire support for a battalion insertion into an area of operations in Shoalwater Bay, the Army's largest training area. We're taking part in the big bi-annual exercise called 'Kangaroo Three'. It's a huge exercise with much of Australia's Army plus Air Force plus US Marines involved.

The CO grabs my arm. He points into the distance and I see that he's looking at a column of smoke. Suddenly the helicopter we're in goes into a spiral... downwards we go. We hit the ground hard and the pilots are yelling: 'get out, get out, now!' We scramble out and are left standing there as the aircraft takes off again.

We stare at each other.

'What the flaming hell happened there?' we say to each other almost simultaneously.

Only a few seconds and we both know. Over our respective radio nets, we hear the call: 'helicopter down, helicopter down' and we understand. A military helicopter pilot's first response is always to go to the aid of the stricken aircraft. That pall of smoke is an aircraft on fire. Our pilots may have to fly casualties out. They can't do that if they are full of passengers.

Who was on the aircraft I wonder? Then I learn that it was a sortie of battalion personnel. I breathe a sigh of relief. It's not that I don't care but at least none of my men will be on board. We wait for a few minutes longer and another helicopter comes to take us back. The exercise is suspended, then cancelled. This is the last day of a three-week exercise anyway, so we are taken back to the airfield from where we will be shuttled back to our base at Lavarack Barracks, near Townsville. The large dusty strip is called Williamson Airfield, and around the edge of it, there are already hundreds of men. They're sitting on their packs or lying under small bushes trying to get some shade. They're all waiting to be assigned to a flight out of this place. I join the waiting throng. My battery isn't here. They're still out in the bush and right now they would be going 'out of action', packing their stores and getting ready to drive the 500 or so kilometres home. So, I sit with my Battery Commander's Party and wait... with the rest of them.

I must've fallen asleep because I feel a hand on my shoulder, shaking me...

'Boss! Boss!' The sound of a worried, urgent voice! It is Mal Booth, one of my Lieutenant FOs.

'They are looking for you. One of our FO parties was on that aircraft. There has been a death and some serious injuries, and the chopper caught fire... and there is no word yet who the casualties are.'

It transpired that the helicopter, which had a mixture of battalion and battery people on it, had lost its tail rotor whilst flying just a few feet above the trees. The pilot had managed to warn his passengers a fraction of a second before the helicopter hit the tops of the trees, flipped over, and smashed to the ground, upside down. On the aircraft was Stevo's FO party, plus a number of other battalion soldiers, one of whom had panicked and undone his seat belt when he heard the warning. He died on impact. The two pilots were alive but injured. One of the door gunners had been injured and was trapped. His helmet was jammed in the fuselage and he was unable to undo the strap. Stevo had walked away from the crash, but without realising it, he had broken his back and he could only manage a few steps before collapsing. Smoke had started to spiral from somewhere and the able survivors, after escaping the aircraft, had come back to rescue the two pilots and the door gunner.

However, the door gunner couldn't be budged. Unable to undo the head strap, Stevo's FO Assistant, a Fijian Bombardier whose name was Bombardier Ratatagia, but who we all called Tracker, had taken action. Tracker was a strong man. He raced around the aircraft and jumped into the fuselage which was ready to explode and burn any second. Whilst the others dragged the pilots to safety, Tracker smashed his fist down on the top of the door gunner's helmet, thereby releasing it from the wreckage. Out to safety popped the door gunner! Snap went two of Tracker's knuckles, along with his wrist and his elbow! Such was the force that he brought to bear! All the passengers and crew were now safe. As they stood back, and before the first rescue helicopter landed, the fuselage exploded into a ball of flames.

They took Stevo and Tracker to the Greenslopes Military Hospital in Brisbane. Tracker's broken bones were fixed but Stevo stayed there and I went to visit him a few weeks later. He had broken his back, badly. He would still be able to walk but would never be the same person again. He was destined for a life in pain. At that stage, when I visited him, he was still unaware of just how serious his back injury was. Lying there on his back, almost unable to move, he cracked jokes about the pretty nurses who would come and wash him, and secretly give him a smoke, and come back on night duty and chat with him. He was an intelligent, handsome, funny and vibrant young man whose life was changed for the worse. I believe the nurses knew better than him, what his fate was going to be. The Army kept him on, in an administrative role, but his work suffered. He was in constant pain. I met him a few times after, in the corridors of Canberra as they say, and we would talk of the good times. Within 15 years of the crash, Stevo died of his injuries.

When we were not exercising and I had time off, Lo and I would pack up the kids, jump into the van and head off to one of the many creeks or rivers that were within easy reach of Townsville. Sometimes we rented the really cheap Army cottages on Magnetic Island. That was some treat!

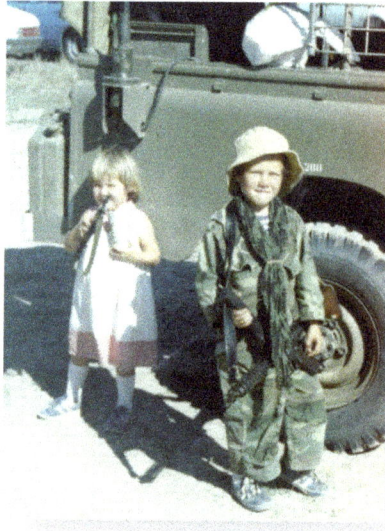

My son Matt (holding a real F1 sub machine gun) and daughter Rae (holding a toy one), at the regimental families' day on High Range Training Area in 1981.
Photo Lo Lans

The kids just loved that water. It can be tricky in North Queensland, avoiding the box jellyfish, but we were well informed of the way in which these deadly jellyfish migrated about, and we knew which beaches and bays to use, and which to avoid. Matt started his schooling in Townsville at the Weir Primary School not far from our house. I can see him now, with his red-brimmed hat on.

The regiment organised a families' day on High Range which was designed to familiarise the families with what their husbands did when they went on exercise in the bush. High Range was only about an hour's drive from Townsville, so the families could easily drive there. The kids were always very excited when I came home from these exercises, because I would bring home bits of my rations packs that would be left over, which fascinated them. Bits like chockie bars and PK chewing gum and biscuits and cans of cheese. If there was a Land Rover delivering me home, they would clamber all over it and try to play with the radios, much to the chagrin of the driver. Now on this families' day they had the chance to see lots of Army vehicles and even ride in them! It was a wonderful day. Robbie was just a baby and Rae was too small to remember. But Matt, well, that was another thing. He came dressed in his camouflage gear with his toy gun (much frowned at these days) and he well remembers the excitement of sitting in the back of an armoured personnel carrier and the noise of the howitzers firing.

On another occasion, we hosted a large group of cadets from Duntroon, the Royal Military College in Canberra. They were in their last year touring all the Army's corps to help them decide which corps they would join. 108 Battery was to treat them to a fire power demonstration where these cadets, all 60 or 70 of them, with compass and maps in hand, had been placed in a large group right behind our OP. From here they would hear all the fire orders broadcast over the radio net, amplified through large speaker, and follow the action. Ahhh, but the CO had not reckoned on the jokers and the characters of 108!

'Hey, boss... let's set them up! Let's see if we can make them run!' one of them said to me. The plot was created. The players readied and lines rehearsed.

For this occasion, the normal calls for fire were broadcast over loudspeakers. The FO ordered the fire and soon 'shot over' was reported from the battery, indicating that the round had been fired and was on its way. Everybody on the OP could hear, but the plans to create chaos were in place. Suddenly, seconds after the report that the round had been fired, the battery radio signaller announced in a panicked voice over the radio saying:

'MISTAKE! ... MISTAKE FIRED! ... TAKE COVER! ... TAKE COVER! ...OVER!'

This was our cue. All of the 108 people on the OP, including me, jumped up and flung themselves on the ground, some over-acting the part, running away yelling 'get down, get down!' Chaos ensued! Cadets started flinging themselves into the bushes, throwing their maps into the air, running away full pelt in all directions! It was a complete rout! Even the CO, who was in on the action, and the college hierarchy standing with him, looked unsure of themselves. I was able to warn the CO that something was going to happen and not to react, but I wasn't game to tell him exactly what! Then, when nothing bad happened and the round went off 'on target' exactly as it was supposed to, the game was up. We rolled about laughing! Dozens of cadets got up, red faced, rummaging around for their gear.

'Bloody gunners, very funny, ha ha!' was heard all about the place.

I looked at the CO to see if I still had a job and yes, I had judged him well. He was smiling! The other dignitaries, after looking at him, also decided to smile! After that little act, I wondered how many cadets would decide upon artillery as their first choice for corps allocation. Well, we wouldn't want any cadets with noses out of joint anyway!

Regimental life was often routine but never boring. The regiment had an officers mess, a sergeant's mess and a soldier's canteen and these frequently held functions. They were places to relax and have a quiet drink with family or friends. In those days, the days before the 'booze bus', we were expected to go and have a drink with the CO, or with the sergeants or soldiers at the end of the working week. This was always something of an irritant to me because, although I like a beer or a wine or even better, a smooth whiskey, I do not like being told *when* to drink. But on the positive side, the messes provided a respite, a place where you could be yourself. On a few occasions, Lo and I brought the little ones along, and they thought it was just great as they always got a coke and a packet of chips. The alcohol was cheaper and the messes provided a bar and a dining room as well as an ante room. Many a dining-in night was had, with Lo on one occasion bringing her parents along. They loved it. The pomp and ceremony of those occasions, with everyone dressed in their finest, was an opportunity to savour in a world where formal 'do's' were fast becoming a thing of the past.

During the second year of my posting, the CO decided that he wanted to test the regiment out on a deployment that would criss-cross Cape York, to practice long range deployments. We were transported in Hercs to a dirt airstrip somewhere up north then drove our convoys hundreds of miles.

Unfortunately, I managed to get sick with pneumonia. This was a surprise to everybody, most of all me, so I was despatched back to Cooktown Hospital to recover. Hospital was something of a grand description of this farm-like building. It had wide, open verandas which contained the patient beds, whilst the nurses' stations, treatment rooms and administrative areas were all in the middle. I was put to bed with a high temperature but in the bed next to me was a very rowdy Aborigine who stole my food, even whilst I was trying to eat it. Chastising him had no effect, so the sister came to me and said:

'Dearie, we are bringing you inside the hospital, otherwise you will not get a moment's sleep, but the only spare room is an air-conditioned one!'

Well that was fine by me. I slept like a log. The next morning the same sister asked me if I slept okay.

'Yes,' I said. 'Very well!'

'That's good,' she said, 'Because this is the morgue. It was the only room spare!'

For the first time, I realised that there were flat stainless-steel tables around me, instead of other beds! Luckily, I had no roommates! After another night in my special air-conditioned room, I was released. The doctor would not allow me to re-join the exercise, so I hitched a ride home to Townsville with a woman who owned and operated, and piloted, her own small plane around north Queensland. Lo was happy. She had me home for a couple of extra unplanned weeks.

We trained to be proficient for day *and* night operations. This included helicopter deployments, where the entire battery was lifted by large helicopters and dropped somewhere forward. Moving a battery by air with all its personnel and ammunition is no mean feat. To do this in daylight was one thing, but to do it in the dark of night, using only minimum lights, was entirely another matter. It takes precise and detailed planning and execution, and can be very dangerous. During such deployments, helicopters, up to three at a time, fly in and out of the gun position, lining up to have their loads hooked up or waiting to be directed to the next drop-off point, all the time using only minimal lighting. It is a noisy, potentially confusing and dangerous business. The SAS practice their night deployments with paired helicopters which fly in side by side, in darkness, landing next to each other simultaneously to disgorge their troops and taking off as quickly as possible. This is to minimise the time that an enemy might have to identify their landing spot. One night not long after my time in 108, two Chinooks at High Range near Townsville, fully laden with SAS troopers, were coming in to land together when their rotors touched. Both aircraft went down in a fiery ball, killing over 30 men. It was a disaster.

On one of those night air-deployment exercises, we were practicing the usual, standard drills. The guns had to be hooked under the helicopter first, then the helicopter would lift the gun off the ground and a pack of ammunition would then be lifted under that. So, in effect the heli would have two loads swinging under it. To prevent 'enemy' from spotting us, the helicopters were not allowed to show excessive lights, thus the landing lights under the helicopter would only be turned on at the very last moment. This would allow the loadmasters to ensure that the loads were being deposited at the right spot and not on someone's head. Dropping six howitzers and six packs of 100 or so rounds of high explosive ammunition packed in boxes, plus a couple of loads of command post stores and equipment at pre-determined spots within a pre-selected area not much bigger than 100m x 100m, in the pitch black of night, is not without its challenges. On this deployment, a troop of three Chinook helicopters was doing the shuffle. I was watching from nearby, not directly involved. Planning and commanding a night deployment is the job of the Battery Captain, not the Battery Commander, who is normally up front with the infantry. For this practice, I sat and watched from a distance.

Then it happened, chaos! As one of the helis came in with a gun and a pack of ammunition underslung, it missed its mark by about 25 metres.

As the heli switched on its lights at the last moment, the mistake it was about to make was seen by the load master. There, illuminated by the landing lights, was a command post tent flying through the air, with three of the operators hanging on trying to keep it grounded, but succeeding only in swinging in the air with the tent. The pilot gunned it. I could hear the strain of the powerful turbines smashing the blades through the air. It's a sound that, once heard is never forgotten; that whack-whacking of metal against air amidst a whirlwind of dust, stones and tree branches! The Chinook, which has two large overhead propellers instead of one, is exceptionally loud and noisy! Luckily, just in time, it began to rise. The tent, complete with soldiers, crashed back to earth, and nobody was injured. It was a close call!

When having to deploy long distances, we often used RAAF Hercules aircraft, also by night. Hercs are large, four-engined, fixed-wing air transport workhorses. On this occasion, we were to rendezvous with the rest of the regiment somewhere near Charters Towers, west of Townsville. Several guns, this time complete with gun tractors, were in the back. I'd been invited to sit up front with the pilots which was an experience in itself! They were flying 'map of the earth', very low—not much above tree level, to avoid radar detection. This was 'seat of your pants' flying, and I was greatly impressed with these RAAF pilots as they flew effortlessly, all the time knowing exactly where they were at an altitude of less than 500 ft. For a big aircraft, this is akin to virtually touching the tree tops!

Suddenly they said 'Okay mate, better go back to the hold, we are approaching the RV.'

'Really?' I said. 'Can't see a bloody thing myself!'

But I did what I was told and buckled in at the back. One sharp, low altitude turn, just one... and down we went, full bore onto the dirt landing strip. The Herc raced along the strip at a seemingly reckless pace until the pilots threw it into a 180-degree turn, jammed on the brakes and stopped, with the engines still belting at landing-come-take-off speed, propellers feathered.

'Out...now,' yelled the Loadmaster, and we unbuckled and ran into the black of night, down the ramp that had been lowered, the drivers getting into their vehicles to drive out with guns in tow.

There is something foreign about being told to run down the back ramp of a Herc which has just landed on a dirt strip in the middle of nowhere, straight into the inky black, all the time hearing and feeling the Herc engines roaring at full throttle ready to take off again, apparently from right above your head! We did the drill. Run 20 metres! Down! Head down, ears and eyes covered! Move your head and bucket loads of fine dust will be inside your ears and eye lids! Within moments, and just as suddenly, the aircraft roars away and takes off.

Then follows a moment of total and complete disorientation. That sensation, when the silence of the black night creeps into your head and replaces the roar, is so stark, so foreign, that it nullifies your every thought or action. It's hard to think straight. Just for a short time, in spite of the briefs and preparatory orders you have in your head, you have no idea of direction or where your people are, and it is difficult to remember what you were supposed to do. Then, slowly, you remember; and then a small light flickers—a signal—somewhere out there in the darkness is a guide waiting for you, and order is restored in your brain! You remember the orders and the sequence, and you get the hell off this strip before the next sortie comes in!

16 108 BATTERY ADVENTURE TRAINING—THE MITCHELL RIVER

The political environment for the military after the Vietnam war was not promising and I firmed in my belief that, in this post-Vietnam era, the Army needed something extra to offer its soldiers. An army needs to train, but train for what? The military was temporarily out of favour with the electorate! Defence budgets were being slashed. Being a professional soldier in that post-Vietnam era was sometimes deflating. The endless repetitive cycle that was the training regime of the day, had the potential to stifle soldiers and unhinge anticipated training outcomes. I witnessed artillery commanders responding by demanding more of the same: more skills training, more weapons training, more collective training... more of everything they were already doing. Then, they wondered why soldiers were leaving the Army in droves?

I believed that there were two reasons. Firstly, Australia's economy was strong and young soldiers wanted to try their hand 'out there in the real world'. Secondly was that they were, to an extent, bored. There was little doubt that the soldiers in 108 Battery were tuned to a high level of technical competence and I came to the conclusion that the best way to ensure that a high level of training was maintained, once achieved, was to distract the soldiers with something entirely out of the ordinary. I wanted to take them away from the hum drum of daily regimental life and challenge them in a way that was different—but complimentary—to their normal military training. They were on a fine edge. I needed to manage that edge. To me, adventure training—seen as frivolous by many senior officers—was the answer. Some thought it, at one extreme, as having a bit of a holiday on a beach somewhere; while at the other extreme, as a privilege for a selected few—mostly officers. To the British Army, the world leader in adventure training, there was no stigma associated with adventure training because they understood the value of it and had it down to a fine art. They sent adventure training expeditions all over the world and set a standard that has never been reached by any other Army.

Adventure training is all about building the individual's confidence and creating team spirit. It's about challenging the individual to achieve personal and team goals that they might not otherwise achieve: to accept, face and overcome dangerous situations that are a threat to their well-being, perhaps a threat to their very life. The trick is to ensure that there is not only perceived but real, calculated danger built into the training. It is imperative that this danger is introduced with a level of risk management that reduces the potential of real loss of life to an acceptable level.

As our Herbert River canoeing expedition proved, conducting training of this nature often brings out qualities in men that normal training does not. This is because, in an adventure training activity, the normal restrictions of rank and position that exist in the military structure, which has the potential to nullify initiative and stifle individuality, don't exist. A carefully executed adventure training activity enables commanders to identify leadership potential by witnessing how their soldiers behave in challenging, sometimes dangerous or unpredictable situations.

I knew that I had many challenges ahead of me, not the least of which was overcoming the stigma, by creating authentic training that the Army would approve, and then melding it into the battery's and regiments' training programmes, which were already quite full. To silence that world of doubters out there, and cut criticism off at the knees, I had to ensure that 108 Battery excelled at the annual Divisional Artillery technical assessment exercises. The challenge was to bring onside not only the hierarchy of the regiment and the Army, but the very soldiers for whom I was designing the training.

It had to be a voluntary activity but I needed the majority of the battery to be there to achieve the team morale building outcomes and there was one other vital ingredient that was absolute. I needed the support of the families. The battery was already out on exercise for many months of the year. This meant that soldiers with families frequently had to leave their wives and children alone for long periods, many weeks at a time. Wives of Army personnel had mostly become very competent at running the affairs of their family. There was little choice, as their husbands were away so much of the time. Now I was asking them to *volunteer* to go away even more. I had to ensure that all the families understood and were happy to let their boyfriend, husband or daddy go away again.

My plan was to conduct two major activities, one in each of the years that I was to be the BC. In the first year, we would do the Mitchell River in far north Queensland, a rarely travelled river flowing through wild, remote Cape York. In the second year, we would do the Franklin River in Tasmania, a river so remote and challenging it had only been travelled by a handful of people. Both activities would require detailed long-term planning, major logistical back-up and many levels of support, both from within and without the regiment.

The Mitchell River flows west out of the Atherton Tablelands, located north-west of Cairns in Queensland, travelling some 750 kilometres across Cape York to the Gulf of Carpentaria. It is crossed by very few roads, even today. The river is fed by seasonal monsoons, collecting its water from the tropical forests west of Cairns. It runs through sclerophyll forests in the central areas and woodlands of Cape York, turning to the savannah in the western plains and then enters the tidal plains that guide the river to its mouth. It only drops about 350m throughout its entire length, so rapids are not a problem. This was an river ideal for open canadian canoes.

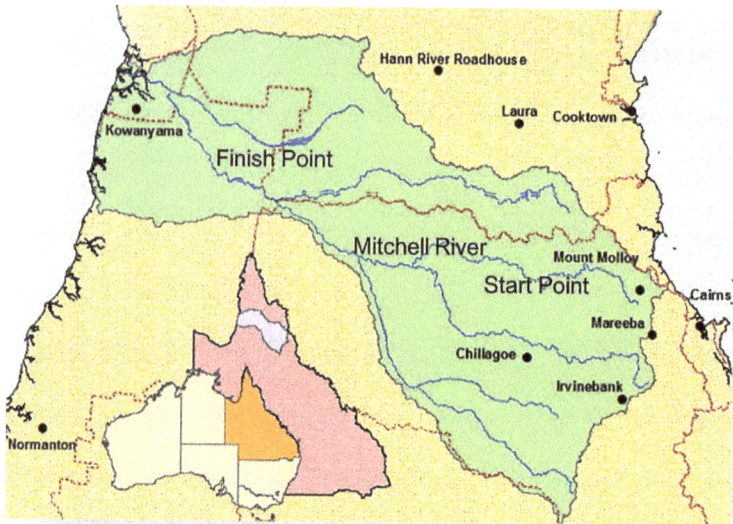

Map: Mitchell River flowing through Cape York Peninsula, North Queensland.
Map Mitchell River Water Management Group Inc with Author's notations

My plan was for the battery to paddle about 400km of the river from near its origin just west of Cairns to a point where it enters the tidal plains. Maximum numbers would take part, and those not able or willing to paddle would be part of the support party, which would be meeting the river group at various rendezvous points. Three weeks on the river was the plan. Rations would be reduced to about half of what a body would need to survive, and the rest of the food would have to be caught by any means possible along the river. A couple of members of the battery had civilian gun licenses so we took two .22mm rifles and shotguns along, as well as ample fishing lines and hooks. Once again, the Army did not own any canoes. Where were we going to source them?

I decided to hand this problem back to the boys in the battery. With about nine months to go before the seasonal rains would likely make the river 'paddleable' again, I called planning meetings and I put it to them. I was well aware of the existence of the Townsville Canoe Club, and that they had moulds for making canoes, and I had discussed with the Canoe Club the possibility of us using their canoe mould and their facilities.

One of the club members suggested that the club needed a new mould and that, if the battery was able to give them a hand making that new mould, we could have the old one for free.

I put this genesis of an idea to the battery and a group of Senior NCOs, led by a Sergeant called Polly Farmer, got together and decided that they would do the job. They organised work parties to go to the club on weekends and in no time at all, the Townsville Canoe Club had their new mould, and we had the old one. Polly Farmer set up an adventure training committee and Norm Wheeler, another of my long-term NCOs, opened a bank account. All the soldiers who wanted to be a part of the trip, and that was most of them, started contributing money to the account. Soon we had enough funds to start buying 100 litre drums of resin, large rolls of fibreglass and sundry canoe making equipment. And that's how it started.

Some of the boys came around to the family house in Kirwan where we made a couple of canoes as a trial run; then we set up the regimental construction facilities back in barracks and in no time at all, the assembly line was pushing out canadian canoes by the dozen, literally. The boys were so pleased with their results that soon soldiers and their families were going out in groups on weekend excursions with their new, proudly self-made canoes: paddling, enjoying picnics and generally getting together. Great for morale!

The day arrived. It was November 1980. We set off on a long journey towards Cairns and the start of the river, or at least that part of the river where we would have enough water to make paddling possible. We were all go! I led a strange looking convoy of Army trucks stacked high with canoes on the road north. But just before we edged out of town, Lo met us at a pre-arranged spot. Jo, our beautiful German Shepherd dog was coming as well! She jumped up into the back of my Land Rover and stuck her head out the back, ears flying, much to the amusement of the blokes in the battery.

We drove all day until, on the road from Cairns to Mount Isa, we crossed the Mitchell River where we set up camp. At this point it was not much more than a creek really.

'Are you sure this piddle actually goes all the way to the sea, boss?' one of the gunnies asked.

'We'll find out, I guess,' came my not overly-confident reply.

I was relying entirely on maps, annual rain charts and word of mouth, and there was not much of that since this was not a river frequented by canoe clubs. The logistics of getting out of the river from some point hundreds of kilometres downstream and finding a way back to civilisation was beyond the reach of most people. In the late 70's the future armadas of 4WDs and campers had not yet explored every corner of this continent.

The next morning, dozens of multi-coloured canoes of the 'Great 108 River Battery Mitchell River Expedition' set off, not in a gloriously impressive float-past, but more in a haphazard conglomerate of total confusion! The canoes varied dramatically in quality. Some of the soldiers had taken great care to create masterpieces, beautifully crafted with glossy gelcoat and polished wooden gunnels. They were proud of their effort. Others were not as carefully crafted, but they all floated.

Gunner John Webb my paddling partner on the Mitchel River, with Joe my German Shepherd. She had her ears back ready to go again after a fierce dunking where she was trapped under the canoe.
Photo Author

I watched the blokes spearing off in all directions, some straight into others, as the procession spectacularly bumped its way along a river so narrow at this point that the tree canopy from either side of the river touched overhead and I wondered if I had made the right decision. Nevertheless, with a supreme show of confidence, I waved goodbye to the road party and we edged our way into the wilderness of Cape York. The river sparkled in the mottled sunshine, free and beautiful, home to an abundance of Aussie creatures, barely visited and rarely crossed. Adventure here we come!

Within days we had left any semblance of civilisation behind, no cattle stations, no roads, no signs of fishermen—just a crystal clear, pristine river. The abundant native fish species in the river was the black bream, otherwise known as the 'sooty grunter'. These fish, not knowing that humans in a canoe spelt danger, swam right up next us. Hooks with silver paper attached, hanging on a line trailing behind the canoe, was an easy way of catching them. The fish would line up to take the bait and be pulled into the boat in rapid succession. Every lunch break, a couple of the boys led by 'Darky',

Gunner Boyd and Sergeant Kewley with the day's lunch bag of fish caught in less than 20 minutes.
Photo Author

my Warrant Officer whose official job when on deployment was 'Battery Guide', would volunteer to go and catch fish whilst others would start the cooking fires. By the time the fires were at the coals stage, the fishermen would be back with 30 or 40 freshly caught fish, enough to feed us all. It only occurs to me now that Darky had some Asian blood in him, giving him a distinct Indo-Malay appearance, hence the nickname 'Darky'. There was no political correctness in those days and nobody took offence.

We feasted on fresh fish roasted over an open fire! Crunchy skin, white flesh, delicious! I doubt if I have ever eaten fish better than that. At night, we would supplement our diet with wallaby, shot by our hunters. We would carve up the wallaby and roast joints over open fires. Jo, my dog, was very happy with all this. I never worried about having to feed her. The soldiers loved her and at dinner time she was always on the receiving end of some good bush meat. One night though, she found a pile of innards that had been left over after someone had gutted a wallaby. She ate the lot without my knowledge. That night, as I was laying in my sleeping bag, I awoke to an incredible stench assaulting my nose. When I opened my eyes, there before me, about six inches from my nose, was a neat pile of guts and other wallaby parts that had been vomited up by Jo.

'Jeez dog, couldn't you have done that somewhere else?' I yelled at her accusingly.

Apparently not, because she barely moved, quite exhausted after her day on the river.

Jo soon settled into life on the river. She became expert at sitting quietly in the centre of the canoe, knowing that if she moved too much she was likely to get a paddle across the back of the head. She quickly understood the rhythms of the canoe and the moment the canoe started to wobble and go faster, down would go her ears and down she would squat, hugging the bottom of the canoe, toenails extended for extra grip. She stuck to me like glue everywhere I went, not letting me out of her sight. One morning, when she was a little late to wake, probably recovering from another wallaby overload, she must have thought that I had deserted her. She jumped up with a start, decided that she had been left behind, and raced towards the bank, ignoring entirely the possibility that I might not have taken off yet. The bank at this point was nice and grassy and about a foot above the river, with a neat custom-made edge, perfectly designed for hopping into a canoe, enabling your first few moments on the river to be nice and dry.

I was actually right there next to her, watching her strange behaviour, ears up and emitting that distinctive German Shepherd whine that is unique to the breed. But she didn't see me! She was in a panic. Racing to the bank and without stopping for a good look, she launched, aiming directly at a canoe that had just pushed out from the bank. It was a magnificent sight, this German Shepherd flying metres through the air, just landing her claws on the edge of the gunnel, where she hung on! Instantly, the canoe flipped. Unceremoniously, she spilled the two unsuspecting paddlers into the water, along with all their gear, which was nice and dry at the start of the paddling day. She swam around to each of them, gave each a good sniff, decided that I was not there, and swam back to the bank, where, upon finding me, she let out a jubilant woof... and sat down! She was happy! But not so the two gunnies who had been tipped in!

'Bloody, hell boss! Can't you keep control of your bloody dog?' That was the censored version of their response. The campfire stories that night were not so polite.

Webby, who was once again my paddling partner, as he had been on the Herbert River the year before, and I were getting pretty proficient at this paddling caper. One day, however, whilst dealing with one of the fast-water races near a small rapid, we managed to paddle an approach line that was completely wrong. As a result, we propelled ourselves straight into the ti-tree growing along the side! We instantly flipped! Immediately the canoe filled with water and was caught, held upside down by the unrelenting force of current pushing it into the bushes. Webby and I were spewed into the maelstrom and swept downstream where, once we recovered, we expected to see Jo. She knew how to deal with these tip-ins and usually swam downstream a bit until she could get to the bank. Not this time. We were frantic! Where was she? How far down would she have been swept? After a panicky search, we raced back to the canoe to hop in, intending to search for her further downstream but, as we turned the canoe over, out popped one very bedraggled dog, with huge eyes and very flat ears. Jo had been caught in the air bubble under the upturned canoe. She could breathe there, so she simply stayed put. In the pitch black, water gushing all around with presumably just her paws on the gunnels. What a dog! I was so relieved... but she was even more relieved! I doubt that any dog would have understood what was happening. From that moment on she stuck even closer to my side.

The weeks came and went and most of the canoes managed to hold together. For those that didn't we used the ubiquitous '100 miles an hour tape'. Aptly named by the gunners, it is the same gaffer tape that muso's use on sound stages. Night after night we camped on sandy banks under clear starry skies, ate bounty from the river and the land, and spun stories about anything and everything. Darky had worked out how to catch the fresh water yabbies, or that is what we thought they were. These were huge, prawn-like things that were much larger than a man's hand. They were easily caught. Once again, the silver paper was employed to lure them near to shore. One person would pull the string with the silver paper and the other would stand by with a fork tied to the end of a stick. Once it was close to shore it would be speared! Later we found out that it was indeed fresh water prawn that we had been eating and that it was plentiful, but uniquely found in the waters of the Mitchell. I wonder if they are still plentiful today?

The daily paddle was hard, but the reward was well worth it. This was as remote and unique as anywhere in Australia—the sort of country explorers like Ludwig Leichhardt might have encountered. The sort of country the Aboriginal people surely thrived in. But now it was empty of people and although in those days large portions of Cape York were under pastoral lease, we didn't see any sign of cattle or sheep. Watching the boys sort out how to make a living on this river, how to survive, deciding who caught and cooked the food, and who distributed it, was greatly satisfying. Admittedly, we were in a land of plenty, but these boys were mostly from cities and in spite of being used to living in the bush as soldiers, they were not accustomed to having to feed themselves off the land as adventurers. There was no Warrant Officer Caterer here to feed them every day! It was actually a relief not to have to rely on the usual 'hard' rations.

On about day 10 or 11, we came across tyre tracks right on the bank, and moments later someone found fishing nets, strung from side to side. We were so far from anyone and anything, even to see vehicle tracks was a surprise. Fishing nets?... In this pristine river? It was obvious that this was an illegal fishing trap and that a vehicle would come, probably regularly, to haul the bounty off to market for a bit of cash on the side. The boys wanted to cut the net into pieces. For a moment I wondered about the consequences of such an act, and it concerned me. Who were we dealing with here? I didn't let them. Perhaps I should have.

Slowly the scenery started to change. The gentle sandy shores and the low eucalypts and other native trees gave way to mangrove varieties and the water was spreading wider. It was obvious we were nearing the ocean, and, although by my reckoning it was still more than 50 kilometres away, we, like Burke and Wills, decided to stop. Going closer to the ocean would mean having to deal with crocodiles. We weren't planning on becoming part of the food chain!

The trip was a stand-out success. We met up with the road party and sat around our final camp fires and re-lived the tales of past fortnight. The cold beer delivered by the road party helped to embellish the stories. The canoes were loaded onto the trucks and we headed home to our families with many memories and tales to tell. This was the second major adventure training exercise that I had conducted with the battery and I was beginning to understand that this adventure training thing was a formula for success.

Franklin River

17 THE FRANKLIN RIVER

'The Last Wild River'. In 1980, that was the rallying cry and what an effective one it was! A worldwide campaign using that slogan, together with Peter Dombrovskis' photo of Rock Island Bend taken on one of the very early river explorations in the late 70's, was to set new standards in conservation of the environment. The Franklin remains one of the most spectacular free-flowing rivers in Australia, indeed, the world. It flows uninhibited through the south-west wilderness of Tasmania; amongst dark mountain ranges and mysterious, sheer gorges; through ancient rarely trodden rainforests; and deep ravines past hidden caves that supported an ancient people who had survived there for more than 20,000 years.

For a distance of just 143 kilometres the river flows unencumbered by anything man made, through country so impassable, so wild, that it had prevented any descent by any kind of watercraft until the 1950s. Even then, it was not travelled again until the mid-seventies. In the early days of colonial settlement, convict inmates of the infamous Sarah Island Penitentiary in Macquarie Harbour were forced to cut Huon pine from the lower reaches and not long after, tough timber getters known as 'piners' dragged their wooden boats to the middle reaches. Early explores, including Lady Franklin, wife of the Governor of the time, and a few surveyors, crossed the river here or there. However, in the last 100 years or so until the middle of the 20th century, the only humans who laid eyes on the river were occasional trappers venturing deep into the south west.

A group of convicts who escaped from Macquarie Harbour would have been the first to see the upper reaches of the river, on their way to Hobart Town, on a trek across the width of Tasmania in the days when no-one had any idea of what lay in the centre of the island. The place was hostile, with little evidence of anything a white man could eat. One convict died from snake bite and the others cannibalised each other until only one survived. Today, apart from a few determined bushwalkers and river rafters, and one road bridge, the catchment remains pristine.

In 1981, however, the Franklin River was under threat because it was marked for damming. The call to save it came from the environment movement led by Bob Brown, the founder of the Tasmanian Wilderness Society (TWS) and who later went on to become a Senator for Tasmania. As was the way of those times, the Tasmanian Government was damming every river that had any potential to create hydroelectricity. This momentum to dam rivers was almost unstoppable because the Hydro Electric Authority, affectionately referred to as 'the Hydro' by Tasmanians, had a ready-made workforce that simply moved from site to site as each river was dammed. Commendable though the creation of hydro-electric power is, it always comes at a cost. According to the environment movement, and a majority of Australians, the cost of losing such a river in such a pristine environment was an unacceptable price to pay.

In 1983, Australian voters forced a Federal Liberal Government to fall, and a subsequent High Court action against the State of Tasmania, initiated by the newly-elected Federal Labor Government, spared the river. It was a controversial decision that shocked Tasmania to its very core. The river would be allowed to flow free forever, central to a World Heritage Area that encompasses magnificent temperate rainforests, ancient pencil pines, plateaus of button grass and peat, and rivers that run wild and free.

However, when I was in Townsville in 1981, that decision had not yet been made, and the fate of the river came to my attention one day, as I was sitting in the lounge in Townsville with Lo, watching a documentary about Tassie on TV. The documentary showed brief glimpses of a wild river in Tasmania. I was hooked! In that brief moment of time, whilst sitting in the stupefying heat and humidity of Far North Queensland, an ineffaceable thought settled in my head. Would it be possible to travel to Tassie with my battery and complete an adventure training exercise there? Why not? This was true adventure. I didn't even know if the river had ever been travelled before. It looked incredibly dangerous. Clearly there were some major hurdles to overcome, not the least of which was how to get my battery from Townsville to Tasmania, thousands of air miles. The equipment: what would we need? Safety: how could I keep my men safe? The rapids: they seemed immense, even in the very short clips I'd seen. How would anyone develop the experience to deal with such conditions?

Every plan starts with an embryo of an idea followed by a single step in the right direction. I started my investigation on two fronts. I rang the Tasmanian Environment Centre seeking local information, and I lined the battery up and floated my plan to the soldiers, asking them:

'Who's interested in coming on an adventure training exercise down a wild river I know little about, all the way in Tasmania? It will be difficult, challenging, even life-threatening'?

Well, that question was quickly answered: almost the entire battery wanted to be a part of it! Instantly, the problem became: who would be left out?

The Tasmanian Environment Centre put me in touch with the Tasmanian Wilderness Society, or TWS, who sent me a film about the river, the one that I had seen on the TV. It turned out that it had only been negotiated a few times and by a select group of men. After the initial successful expedition led by John Dean and John Hawkins in the 1950s, Hawkins continued to explore the river over the next two decades, taking others with him including Olegas

Truchanas (the environmental champion of south west Tasmania in the 1950s and 1960s) and Dave Serle, another of the tough pioneers of Tassie's wild rivers. Then in 1977 and again a year or so later, Bob Brown descended the river. On the second descent, he and his paddling partner Kevin Kiernan, discovered a cave on the lower reaches of the river that, on investigation, turned out to be the lowest latitude at which man survived during the last ice age. Bob Brown filmed the descent and made notes about campsites, where the major rapids are, and how to negotiate or portage those rapids. He was using this film to publicise the river as much as possible so that the world would become aware of what might be lost if the river was dammed.

We had about a year to plan for the trip. Whenever I mentioned it to anyone outside of the battery, the standard response was that it couldn't be done: the logistics, the cost, too hard, too dangerous, too far! Some said: 'why don't you just go and camp on a beach with your men and do some fishing?' Stupid statement! The mystique of that wild river had given me an embryo of an idea, and I'd set that idea free. I had made it public! There was no way that I, or indeed my blokes, were going to let this one go. One of my Lieutenants, Mal Booth, was also hooked and, being a national-standard tri-athlete, said he would run the fitness training. Harry Pregnell, Norm Wheeler and Vic Shields, who were senior NCOs in the battery, and 'Tracker' Ratatagia, Alan Ward, Barry Kyrwood and Ian Biggs, who were junior NCOs, understood this once-in-a-lifetime opportunity and wanted a piece of it. These blokes were keen to become my core planning group.

The priority was to gain support. Not support on the river, but support from the Army—my regiment, the brigade, the corps of Artillery and Army Headquarters. A formidable list! Then there was support from the families. Adventure training is a voluntary activity and without support from home, none of the married men would want to come. I'd quickly determined that the best time to raft the river is when the winter river levels begin to subside. Tassie's west coast winter rains can amount to three and a half metres. The most opportune time would be December. This was a problem. To accommodate the regimental training timetable, the window of opportunity for the trip might carry over into Christmas. What would the families think of that? The trip would definitely interfere with the lives of my soldiers, especially the dads among them, who would prefer not to be away for Christmas. The formal approvals would also be difficult. I anticipated that the naysayers in Army Headquarters, of which there were plenty, would do their best to ensure that the activity was ridiculed as 'too dangerous; a waste of money and time, not the sort of thing soldiers do'.

Once again, I thought: 'let's start where I know I can count on support... my CO!' So off I went to see my boss, Lieutenant Colonel McGuiness. The response from that ever-supportive man was entirely predictable. He was so helpful that he volunteered to see the Brigade Commander on my behalf and, incredibly, he would even have a word with General Drabsch, the Divisional Commander. Right, so that was a good start! I also needed a tick of approval from Army Headquarters, or Army Office as it was called at the time, and to have the support of the Brigade and Divisional Commanders would be a game changer. But I still wasn't confident, as no adventure training on this scale had ever been done before. Nevertheless, after an agonising wait of some months, I received my approval, but it was conditional. The activity could go ahead with two major requirements: I was to do a personal reconnaissance of the river; and there would have to be air casualty evacuation support in the form of helicopters. The personal reconnaissance was no problem. I rang the staff at Anglesea Barracks in Hobart and made arrangements. But the helicopter support might be more difficult. I had a mate at Army Headquarters, Major Bob Newton, who interceded on my behalf by imposing these tough conditions, including the potential show-stopping requirement for helicopter support. His action circumvented the almost certain and immediate refusal of my request by some of the overzealous and risk-averse senior officers.

I had to pay my own way to Hobart to meet up with the staff at Anglesea Barracks to do a reconnaissance of the river, as far as the Franklin would allow, anyway. The staff at the Army Regional Headquarters in Anglesea Barracks in Hobart were just as amazed as everybody else that I should want to go down this river with a bunch of soldiers. However, to their credit, one of the Majors on the staff, who was an avid bushwalker and knew a little of the south west, took me in his Land Rover to the site of the proposed Hydro dam on the Franklin. That was the only place that I actually viewed the river. My recce was done.

In my heart, I knew that air support was going to be difficult. Army choppers were always in great demand, and this was a request for their use in December. What chance did I have? But again, the river gods were on my side because, as it turned out, the Army Aviation Corps was also looking for activities to liven up the daily grind of predictable flying training, and the Commander of 161 Aviation Reconnaissance Squadron in Sydney rang me:

'What a great idea mate!' he said. 'I'll give you a fully supported troop. All yours!'

I nearly fell over.

A fully supported troop comprises three Army Light Observation Helicopters, known in the Army as 'Kiowas', together with flight crews, back-up flight crews and maintenance crews. He thought this was a fabulous chance for some of his men to see Tassie and fly the 'wild west' of the state.

Everything remained 'go!'

How to deal with the families was where Harry Pregnell and the core planners came in.

'We'll have to train somewhere, won't we? Why not take the families along when we do the training at a river somewhere here in north Queensland? That'll make them feel a part of the whole thing! We can set up a camp for the families near the river where the littlies will be able to play while the boys go and train!' said Harry, with that inimitable half smile.

A light bulb exploded in my head. What an inspiration Harry was, with his indomitable spirit and buckets of common sense.

As soon as our rubber inflatable rafts or 'rubber duckies' (as they had already been affectionately dubbed by Bob Brown) arrived, we engaged on an ambitious training plan. All the paddlers and support crew going to Tassie, plus a large contingent of the battery's cooks, drivers, signallers and sundry bodies, set up a large camp on the banks of the Herbert River near Ingham, just north of Townsville. The lower Herbert, well below Blencoe Falls where we had exited the river two years ago, with its fast water but gentle introductory rapids, was a good river on which to train. With families, including children in tow, we planned to conduct weekend training sessions that allowed us to have fun with our families and be serious at the same time. This way the paddlers were able to spend time familiarising themselves with the clumsy little crafts (that seemed to have a mind of their own as soon as they hit any rapids) whilst simultaneously playing with their children and having fun. What a stroke of genius! Once gain the CO smiled his approval.

'Good for morale Ben. Just don't kill anyone!' he said.

'I wasn't planning to, Sir', I replied, sounding more confident than I was feeling.

The training camps worked! The plan was executed to perfection and I had to do nothing at all. The NCOs put it into action. At night, there were big camp fires and a few of the blokes pulled out their guitars and sang songs well into the night. Three times the battery did this.

There was not a man that pulled out because of family reasons.

Another major impediment was the sheer cost. Almost every bit of equipment was 'non-Army issue'. In addition to the rafts and paddles, we needed life jackets and helmets, wetsuits, river shoes, sleeping bags, dry bags and drums, wet weather gear, cold weather gear, quality tents (not just Army issue 'hutchies') and sleeping mats, and lots of other items where the Army-issue gear was not suitable or good enough. Then there were things like cameras.

Who was going to be mad enough to take their special camera down a river like the Franklin, where the chance of drowning a camera was a sure bet? The idea of sponsorship, that is, asking commercial companies to provide equipment in return for advertising and 'bragging' rights, was in its infancy, particularly for adventure activities. But it was worth a try, so I devised a plan to take along a professional photographer and a 'connected' writer: one who had contacts that would give him a shoe-in to get published in the major newspapers. With this plan I would write to manufacturers of outdoors equipment and offer them publicity in return for their equipment, free or at a reduced cost. I also arranged for the ABC to do a television documentary about our training adventure.

It all worked well. The Franklin was frequently in the news. It was shaping up to be a world-wide conservation issue. An Army group doing an Army expedition on the Franklin was a great opportunity for the outdoor gear manufacturers to exhibit their stuff! I scored! We were provided with most of our equipment for free, or at least at cost. Shamelessly I hit up my brother in law, the Marketing Manager of Minolta, for SLR cameras to be used by the professional photographer. Not only did he readily agree, he offered me additional Minolta waterproof cameras that had just been released on the market. This opened the opportunity for me to go to Kodak for 35mm film. They also said yes and obliged by giving me over 400 rolls of print and slide film. In addition to Minolta and Kodak as major sponsors, we were also given Featherdown sleeping bags, Primus tents and cooking equipment, Sportscraft Wetsuits, Wilderness Equipment Gore-tex coats, three sets of Hallmark Z-Kote coats and pants, Damart thermal underwear, canvas backpacks, Blundstone boots and other bits and pieces.

With excited anticipation, we watched the gear arrive bit by bit. In the hot tropics of Townsville, we steadily built up our pile of cold weather, white-water river adventuring equipment, generously provided by our many sponsors. It occurred to me that the Army might not see this as such a good idea, as the rules were very much against Army promoting selected products through sponsorship. Reluctantly and, deliberately vague, I asked Army Headquarters for permission and to my surprise, it was given, once again with qualifications. Sponsorship was permitted within a limited set of guidelines.

Part of the team practising at the Barron River in North Queensland. From the left Tracker Ratatagia (who shortly after this was taken was involved in a helicopter accident whilst on exercise in Shoalwater Bay Training Area), Bombardier Ian Biggs, Bombardier John Webb, Bombardier Alan Ward, Major Ben Lans, Gunner Graham Jorgensen, Corporal Mark Lowrie and Captain Ian Johnson (who was leader of Group Two).
Photo Herbert Van Daalen

The guidelines were expressed in a manner that I interpreted as vague, so vague in fact, that I saw it as a loop hole to exploit. Limited sponsorship was not what I was planning. I already had massive sponsorship! I put that problem aside as an issue to deal with later. After the trip, I would face any consequences.

The final training run was to be held on the Barron River west of Cairns. The Barron is a noteworthy white-water paddling destination, fed by the Barron Power Station on the edge of the Atherton Tablelands. The power station conveniently fired up its turbines every day at 3pm, filling the river with icy water from the bottom of the reservoir and supplying all the rapids in the lower Barron. The freezing water took our breath away but was good practice for south west Tassie's cold conditions. Perfect for training.

The training run was another one of those family affairs, with many of the battery members taking their families along. This time Lo and the kids, as well as Lo's mum, Gwen, wanted to come. It was also the opportunity to 'blood' the professional photographer, Herbert Van Daalen, who happened to be my nephew. Herbert had just completed a Technical College photographer's course and was technically qualified. I worried he was a bit young, though. He was younger than most of my soldiers, and I worried that it might be a little intimidating for him to be thrown in with a bunch of hardened Army men. How would they treat him? Was Herbie tough enough? When I had asked him to come along, he agreed immediately, so I was confident he understood the challenge. Nevertheless, I determined not to show him any favouritism. Like it or not, he was going to do a lot of maturing in a very short time. I need not have worried. Herb was up to the task.

Coming from Sydney, Gwen arrived at the Barron before Lo, the kids and me. I had made arrangements for her to be picked up by my battery boys and she was warmly welcomed. So warmly in fact, that Tracker, the big Fijian with the million-dollar smile, talked her into coming along on a practice run. Tracker was still fit at this stage and was on the list to come on the exercise but was later injured in a Shoalwater Bay helicopter smash and could not undertake the Franklin. Tracker didn't have a lot of persuading to do. Of course Gwen would go! She was up for anything at all, especially when a handsome Fijian man invited her. She fell in love on the spot, just as her daughter Lo had done when she first met Tracker.

They set of down the river and apparently, all went well until suddenly a huge wall of water caught them from behind and propelled them through the remaining rapids at great speed. Tracker was unaware that the gates would be opened at three every day, turning the mild river into a maelstrom of foaming white water! He told me later he was suddenly keenly aware that an unauthorised dunking of his boss' mother in law was not the best career move! Unfazed, he handled the sudden increase in water strength with great expertise and more than a little luck. In the meantime, Gwen had no idea there was any problem. When Lo and I and the family arrived, we were just in time to see Gwen alighting the raft, with big eyes full of excitement and wonder at this brave soldier who had just taken her through these amazing rapids! They both had big stories to tell! Tracker had kept her safe. What could I say?

Sponsorship equipment was sorted: now we needed to sort publicity. Mal Booth designed a special logo featuring the battery 'mascot', the little red devil that is normally seen on our battery shield, riding on the back of a 105mm artillery shell as it hurtles itself though the air. For our Franklin version of the logo, the little devil was sitting in a rubber ducky holding a double paddle. I used that logo on our letterheads to the sponsors, on our T-shirts that we had made, and on everything that needed some point of recognition. I named the trip 'Exercise Olegas Truchanas', in honour of the man who was a pioneer in exploring Tassie's south west rivers. He drowned in the sixties whilst attempting to take photographs to use in his endless quest to save Tassie's south west rivers from being dammed.

We managed a couple of articles in the Army Newspaper during November, and I made arrangements with Wild Magazine's editor, Chris Baxter, to provide him with an account of the trip for publication. He obliged by announcing in Wild that our trip was about to commence and that there would be an update to follow. Our attempt at filming the exercise did not work out unfortunately. I had agreed that the ABC would come and film us to make a documentary as part of their series, 'A Big Country'. However, there seemed to be little planning going on, despite my calls to the ABC reminding them. Their answer was always 'Don't worry, it will be fine'. But I knew it would take more than just re-assurances like that, so I lost confidence in the ability of the ABC crews to accompany us or, using the helicopters, to even join us at various points along the river. The director seemed not to understand the difficulties that filming might present. By this time, Captain Ian Johnson who had volunteered to do the trip, had suggested a film company that he knew, which had some experience in filming adventure activities. So I cancelled the ABC and we flew to somewhere on Cape York, once again utilising Army helicopters, to chat with another film producer, who was quite excited and quickly agreed. Unfortunately, this producer also underestimated the effort and when he realised that his crew would be subjected to carting lots of expensive camera equipment and film down the river in rafts, and that they might actually fall in and perhaps even ruin a camera, he also pulled out. This left us without a film to be made, something that I had promised the sponsors. I informed them all, right at the last minute. After all, they were unlikely to pull their support with only a few days to go before the start. In retrospect, I was glad that there was no film to worry about. As it turned out, we had enough on our hands without the worry of a film crew as passengers!

On the 26th of November 1981, the day of departure arrived. In Army lingo, we were approaching the countdown to 'H Hour', the time at which we would cross the start line. The entire party was to be transported on a RAAF Hercules Transport, all the way down the east coast of Australia to Hobart. This was a long way to travel on a 'Herc', where the principal in-flight entertainment was watching some poor individual try to use the lavatory, comprising a toilet basin suspended away from the aircraft cabin wall, whilst trying to keep the flimsy, inadequate curtain closed to maintain at least some dignity. Usually this failed, and it was fortunate that the noise inside a Herc is deafening, drowning out any sound effects of the poor individual on the toilet. In those times before iTunes and when a Walkman was new age, ear plugs and books were the order of the day!'

The Franklin River expedition, as it was being dubbed by the sponsors and the small amount of press publicity we began to receive, comprised: three groups of six paddlers (each in an individual rubber raft); a support party, made up of logistic back-up personnel, Army medics as back up to the river medic with me in the leading group; the helicopter support group of three aircraft and support elements, provided by 161 Aviation Squadron and the battery signallers, who would establish communications. The communications plan included a base station at Strathgordon and other communication sites as determined by need and terrain accessibility. Entirely separate to the river trip, there was a walking group of six, which planned to walk the Overland Track. Some 40 battery personnel left Townsville, plus the Regimental Padre. The characters in the battery joked that the Padre came along to administer the last rites to any drowned gunners. That, plus the fact that the large plastic bags we took for emergency body warming, which the blokes soon dubbed 'body bags', had some of the less confident and inexperienced members worried! Actually, all the Padre really wanted to do was walk the Overland Track!

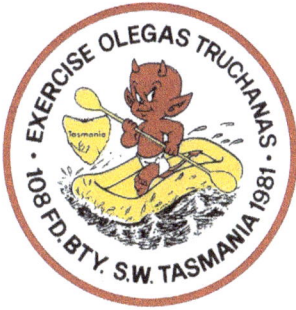

108 Battery special logo designed to provide to sponsors and identify everything to do with the exercise.

We wore special T-shirts on the RAAF Herc aircraft and that nearly brought us unstuck. The Loadmaster on the Sydney to Melbourne leg was going to throw us off the aircraft because we were not in uniform. The rule was military uniform for all military members on a RAAF transport. He became very serious and very threatening, clearly imbued by his importance as Loadmaster with nothing to load but a bunch of Army blokes going on a 'swan' as he saw it. I saw what was happening and as I started to make my way forward, Tracker quietly signalled for me to stay put. I did as I was told by my Bombardier, and after a few minutes the Loadmaster waved us on.

'What did you say to him Tracker?' I asked.

'Not much boss...'

He never said. Tracker was a charming and imposing Islander. The T-shirts didn't have any rank displayed. For all I know he had promoted himself to be the commander, because from that moment on the Loadmaster treated him with deference and totally ignored me! I took note! Tracker's got initiative!

We settled into Anglesea Barracks right in the central part of Hobart, which in itself caused quite a ruckus. Sleepy Hobart had not seen the likes of a team like us, and many a surprised 'Hobartian' eye was cast over us, as we jogged the streets of the CBD in formation... 'double time boys, lef-ri... lef-ri, heads up, breathe deep!' as Mal's voice penetrated the early morning peak hour. The staff at Anglesea treated us well, even gave us the use of a number of Army cars. In those days, all Army cars were painted green 'olive drab', easily identifiable as military. At night we hit the town, using the Army vehicles to get around and the only complaint we received was from the Commander of Anglesea. He asked us to, 'please not park the Army vehicles right next to the door of the Wrest Point Casino!'

'Ah, okay, sorry Sir,' came my very deferential reply!

My original plan was to send the three groups down three different rivers, all meeting up with the Franklin. My group would start on the Collingwood River, which joins the Franklin within the first ten or so kilometres. This was the route established by Bob Brown. A second group, led by Captain Ian Johnson, would walk with their equipment along a high ridge line leading south from the Lyell Highway to the headwaters of the Jane River, then negotiate the Jane until it meets the Franklin. I had researched a track that was used by trappers up until the 60s and thought it would be quite okay to use. The third group, led by Lieutenant Mal Booth, would head cross country for the source of the Denison River, which eventually flows into the Gordon, and paddle their way to the Gordon-Franklin junction. Once again there were signs of old tracks on the maps.

This three pronged approach was planned in the absence of any real first-hand knowledge of the Tasmanian wilderness. The information I had gleaned during the recce was not sufficient to give me real facts about walking and carrying equipment through Tassie's famed south-west. The plan would need updating and we were now told tracks that I thought existed were in fact narrow pathways crudely hacked though almost impenetrable rainforest, unused for decades and entirely grown over. We would employ the back-up plan. Now all three groups would do the Collingwood-Franklin route, departures separated by half a day. My group would be first, Mal's second, and Ian's third. We wanted to keep a distance between the groups because the camping opportunities were few and very tight. Bob Brown had created them on his trips and there was a desire not to cut new camp sites with abandon.

Just before leaving Hobart, I met Bob Brown in a pub next to the barracks. I tried to get him a lift in our helicopters but was refused permission by Army Headquarters. Bob, as a civvy without anything to do with the exercise, was not authorised. I was immediately impressed with this humble but determined man. I was lucky enough to catch him a few more times over the years, including during my time as President of the Queensland Branch of TWS, when we sat near each other during the famous High Court verdict that saved the Franklin. In later years, he always remembered that first time we met in the pub in those early days. With all the work that he did and the many people that he met in his efforts to save the environment and establish the Greens party, not to mention his tenure as Senator, I was astounded that he remembered me at all.

Early on the morning of 25th November 1981, the boys pulled out of Anglesea Barracks in an Army bus, heading for the Lyell Highway and the Collingwood River bridge, our start point. A last chance to stock up on Mars bars and other goodies was seized at the lone service station in Derwent Bridge, where, to the amazement of a startled Tassie shopkeeper, the battery boys cleaned up most of the shop's stock of ice creams and chips and sugary sweets. A long association with a wonderful wild river was about to begin.

My group comprised myself as Exercise Officer in Command (OIC) and Group Leader, plus:

Sergeant Vic Shields
Bombardier Alan Ward
Lance Bombardier Ian Biggs
Corporal Mark Lowrie (Medic)

Mr Don Johnson (Journalist/writer)

Mr Herbert Van Daalen (Photographer)

Group two comprised:

Lieutenant Mal Booth (Group Leader)

Sergeant Norm Wheeler

Gunner Graham Jorgensen

Gunner Steve Boyd

Gunner John Webb

Gunner Tony White

Group three comprised:

Captain Ian Johnson (Group Leader)

Sergeant Peter Nolen

Bombardier Bill Nettleton

Bombardier Norm Hannan

Lance Bombardier Barry Kyrwood

Gunner Tony Stocker

The support group was led by Sergeant Harry Pregnell. In operational life he was my assistant, responsible for managing my non-tactical affairs whilst out on operations. Harry was the kind of 'jack of all trades' who could do anything. The perfect man to trust whilst I was doing my job organising the Artillery support to the Infantry. But here on this exercise he was the 'go to' man for everything.

In spite of being the OIC, I would not be making any minute to minute decisions about the river groups led by Ian Johnson and Mal Booth, quite simply because of the physical separation. We'd spent the last day before departure going through the safety plans and procedures and I had briefed all the members on their responsibilities, including the pilots. Harry would be responsible for taking the initiative should communications fail completely, or the weather turn bad, or some unforeseen incident occurred. He would deploy the helicopters when he believed it to be necessary. This was a huge responsibility for him, but I knew I had the right man to look after us when we were away on the river. Once on the Franklin, apart from moment-to-moment decisions concerning my own group, I would have limited ability to control the overall exercise. I seemed to recall that American servicemen are often depicted saying 'in God we trust' before going into battle. For us it was 'in Harry we trust!'

Each group leader was responsible for the group's 'go or no-go' to negotiate rapids; how and where to place safety people; whether or not to risk any rise in river heights should it rain. Sometimes, when I thought about what we were heading into, I'd break out in a cold sweat. There were many risks that were known. For these we could plan. But for the risks that were as yet unknown, we couldn't plan. You can only prepare for what you know lies ahead. There is no substitute for knowledge... knowledge of the terrain, the unpredictable weather, the river in all its moods. In spite of the training we had done, and the fitness of the men, I knew we would never be fully ready.

Did I understand fully what I was wanting a bunch of gunners, not experienced river men, to do?

We were setting off to have an adventure in a place that was more remote, far wilder and more hostile than anything we had seen before, in water that was dangerously cold, and on a river that was an unpredictable one-way ride: once committed there would be no going back! In the decades after we completed our trip, many a life has been lost in some of that river's notorious, treacherous rapids. Small mistakes, momentary lapses of concentration, even under the guidance of experienced river guides, have led to quick, unexpected deaths. But we didn't know this yet. Here we were, about to set off into the unknown, this river that could change from being a serene and beautiful wonderland to a raging, unforgiving killer in a matter of hours. Were we good enough to deal with it?

The chance of rescue in the event of a casualty was dependant on achieving radio contact. We had some pretty good signallers, but making a radio set that has a range of only 25km, transmit out of a deep ravine, bend its radio waves around some of Australia's most rugged mountain ranges, and talk to a base station that was more than 75km away, was a big ask.

Nevertheless, an element of the unknown is a vital ingredient to successful adventure training. How else do you bring out the initiative in men, make established older leaders stand up and younger ones stand out? We were as trained and equipped as we could be! Or so I thought.

Our food was planned by a nutritionist. We had more food than we knew what to do with, courtesy of the advice provided us by the Army Food Science Establishment at Scottsdale in north eastern Tasmania. The hardest choice was what to do with it all: where to pack it. As was the way, all the food was dry food. But in the end each member packed what they wanted in little plastic bags tied off with rubber bands and stashed here and there. The odd flask of rum—considered a necessity—was also packed. Bars of chocolate were stuffed wherever they could be fitted, as we were all well aware how good a few pieces of 'choccie" tastes at the end of a hard day.

Each person paddled with a large drum and a backpack jammed in front of his legs. The drum contained the 'dry 'essentials. Anything that could get wet was carried in the backpack. An inflatable li-lo was used as the floor of the raft, an essential item for comfortable sleep that doubled to keep your backside from being smashed on the rocks.

Instead of driving to the start point with the boys, I took Herbert with me in the morning and we used the helicopter to reccie the whole area from start to finish before joining the group. This is what I recorded in my journal:

> The (aerial) reconnaissance of the Franklin was interesting and yet I was reluctant to do it as it took away some of the mystique. Frenchmans Cap can only be described as breathtaking. I hope the weather holds. I met the boys on the bus at Derwent Bridge and they were anxious to hear what the state of the river was. I think they feared a report of low water again, as was the case with the Jane and the Denison. Yes, the Franklin is low and challenging.

> Journal Entry, Ex Olegas Truchanas, 26 Nov 80

18 THE UPPER FRANKLIN

As well as being a place that promises adventure, and potentially misadventure, the Franklin is a uniquely beautiful place. The upper part of the Franklin is generally considered to be that stretch from the junction of the Collingwood River to the Great Ravine, some two or three days' travel by raft, depending on how fast the rafters wish to travel.

This is how Don Johnson, the writer come freelance journalist, who I invited to come along to write and publish stories about the trip, described the river:

> The winds that lash south-west Tasmania have a fetch of half the globe. At the end of their journey across the Southern Ocean, they deluge the central highlands and the high bare slopes of button grass funnel the water into the mountain tarns, whence it finds its way back to the sea, through rivers which carve huge gorges and ravines in their courses.
>
> The Franklin is one of the few undisturbed areas remaining from the spacious days, a wilderness where man may, at his mischance, perish. It is the last place in Australia where our species has not left its distinctive spoor, for the area is so remote and so rough that it would require the combined assault of our ant's nest technology to maim it. Until then the Franklin remains a wilderness, attracting

those who find satisfaction in exploring the natural surroundings, unaided by the combined weight of society.

Don Johnson, *The Weekend Australian*, March 6-7 1982

To maximise the distance between groups, my group headed downriver almost as soon as we were dropped at the bridge on the Collingwood River. We understood the water level to be low and were not surprised to find the rafts dragging their bottoms over submerged logs. Some of these were enormous and we soon found out that shore portage was prevented by heavy vegetation which grew deep and strong, right to and overhanging the river's edge. Straight down the middle, over the obstacles, was quickest. Snags such as sharp rocks or tree branches were everywhere.

As we tried to launch from below the bridge over the Collingwood, Mark Lowrie immediately managed to put a hole in his 'virgin' raft, by smacking it down on a steel peg, a left-over from the bridge construction days. We were hopeful that there would be no further man-made obstacles like that, apart from where the Hydro had already made their mark at Propsting Gorge downstream, the place from where I had seen the river on my recce. This is how Don Johnson described the first few hours:

> The blue-grey of exposed quartzite lends a deceptive haze of distance to the landscape. Even a few miles downstream from the Lyell Highway, the sense of isolation is absolute, spiced with the awareness that the only possible direction is downriver, and the only course of survival is by one's own efforts, aided by such assistance as one's companions can afford.

Don Johnson, *The Weekend Australian*, March 6-7 1982

We reached the junction of the Collingwood and Franklin Rivers without mishap and camped the first night amongst trees and sounds and smells that were not yet familiar. This forest was ancient and almost undisturbed, apart from a few visits by the early piners who were after the Huon pines. These were my thoughts on the first camp:

We sit around a smoky fire. Ian Biggs plays his guitar and some of the boys are singing along. The river burbles and gurgles and pops in the background. A reassuring sound that will no doubt become more familiar as we penetrate deeper into the heart and soul of this river. There is a sense of achievement at having made it to the river at all, to the wild south west of Tassie, and an immense sense of anticipation of the unspent days ahead. For so long have we been planning this and here we are. Tents set up in a haphazard fashion, rafts pulled up out of the water and made safe, wet gear laid out to dry and warm clothes on. Each member of the group has his world reduced to: a small, somewhat dodgy four-person raft (four-person by Taiwanese standards, where the rafts are made) but actually suitable only for one; a double-bladed home-made paddle; a waterproof plastic drum containing sleeping gear, warm clothes and essentials like rations, cameras and film; and a pack. The pack is lined with plastic bags to reduce the water intake when the raft turns turtle, containing non-essential dry items like sleeping mats, spare boots for walking out, and a two-man tent. This is Wednesday 26 November 1981, the first day on the river.

Journal Entry, Ben Lans Ex Olegas Truchanas, 26 Nov 81

The next morning, we set off under sunny skies and almost immediately the scenery changed. The dark, imposing forests that pushed right to the water's edge along the Collingwood gave way to soft leafy surrounds, just as thick, but a little less intimidating. The strong southern sunlight piercing the clean, crisp Tasmanian air created a spectacularly beautiful scene; the sparkling tannin-stained water resembling the colour of whiskey contrasting with the blue sky overhead. On both sides, behind the leafy edges which faced the sun, the forest was deep and almost impenetrable. Overhanging the river banks was the occasional old, but mostly young, Huon pine.

We drifted on the gentle current and let the magic sink in for hours, each of us deep in our own thoughts, until we came to a slight widening of the river. Here, a gigantic log lay jammed across the river leaving only a small channel for the water to tumble between two boulders: between them squeezed the entire river's worth of water into a chute not more than a few metres wide! This sharply increased the velocity and pushed the water directly into and under an overhanging rock shelf located at the end of the chute, which in turn caused the water to 'boil' in place and repeatedly turn upon itself in another attempt to pass through the obstacle, before eventually exploding into a pool of bubbly swirls and eddies. This was a serious threat. I was sure that this was the place the Dean and Hawkins party, came to grief in 1952. From here they had to walk out, stashing two canoes for a later return. By the look of the surrounding cliffs and deep rainforest, that was not something that I fancied doing.

John Dean died in 2012. He was a determined white-water adventurer who, in the 1940s and 50s, with his friend Hawkins, used to hitch rides on the Burnie to Queenstown freight train and persuade the train driver to let them off as the train crossed such rivers as the Pieman and the King. They would offload their rudimentary rafts based on inflated tubes and, sometimes in bare feet, travel the rivers, only to have to walk many miles along the west coast beaches to return to civilisation.

We managed the 'Log Jam' as it had been named and proceeded to the next tricky little rapid named 'Nasty Notch'. It was here that Biggsy suffered his first ignominious dunking. Instead of his raft punching its way through the rapid, the river hydraulics forced it to 'high-side', quickly dumping it full of water and flipping it on its edge, tossing Biggsy out and tumbling him end over end, with the raft arriving at the bottom of the rapid upside down. The final moment of indignation came when the raft bumped into the back of his surfacing head!

'Nice one mate'. 'Well done'. 'Stylish!', came the comments from the rest of us. Having recovered his gear and his self-esteem, he decided, with much arm-waving and gesticulating, to direct all of us, one at a time to 'go

Bombardier Ian Biggs, playing his guitar around a smoky fire.
Photo Herbert Van Daalen

right', which, somewhat stupidly, we all did! Had we not all watched him get smashed? But like sheep we followed his directions, resulting in an unceremonious smashing for each of us. When we asked him why he had directed each of us to a guaranteed swim, he smiled and said he didn't want us to miss out on the experience he'd just had. For the rest of the trip we dubbed him the 'go right' man and refused to heed his directions.

I was planning to camp the second night at the Irenabyss, a place so-named by Bob Brown. The Irenabyss is preceded by a seriously long rapid that proved to be the downfall of some of the boys, causing them to enter the Irenabyss chasm well separated from their rafts. But what a glorious place it turned out to be! Deep, inky-black water, utterly still, with high cliffs on both sides, and a friendly, sunny campsite at the end. Such a contrast! Our navigation of this gorge involved a noisy, roller-coaster ride over and around shiny rocks that seemed to reach out to grab us amidst water pounding against the sides of our craft and the gorge, slapping into our faces, followed by complete silence, as if the noise had been swallowed. It was from here that we intended to climb the Cap. The second night—the second fire. Our campfire was becoming an important routine. A way to get warm after a cold day on the river, a chance to dry clothes, dry the li-los, cook dinner and talk about the day's heroics, or wipe-outs. Usually both!

Unfamiliar bird calls, accompanied by the now familiar soft burble of the river, woke us the next morning. I looked for the sun but it takes a while for the sun to reach the bottom of these ravines that carry the Franklin, and we were not even in the deepest ravine yet. The small bits of sky I could see looked dark and foreboding. We secured the rafts well away from the river in case of rain and rising river levels and packed quickly and lightly for the climb. Frenchmans Cap, at 1446 metres, is a prominent peak that can be seen from Macquarie Harbour on the west coast, and beyond—well out to sea.

It was named before Van Diemen's Land was settled, when the early mariners used it as a navigational aid. The first European to have ascended the peak was James Sprent, in 1846, who establish a trig point there. Alexander Pearce led a group of convicts this way on their escape from Sarah Island to Hobart Town and his descriptions of their escape indicated they followed a route along the base of the mountain, between it and the Franklin River. This was given credibility when bushwalkers found a set of convict irons near the Loddon River, which has Frenchmans Cap as its source—certain evidence that the convict party had passed along the base of the Cap after probably crossing the Franklin somewhere near a major rapid that is now called the Churn at the head of the Great Ravine.

We packed light. The first hour was a slog: the rainforest was thick with little room between the trees to allow a body carrying a pack to wedge its way through. The slope was so steep that we needed to use tree branches and roots to drag ourselves up but unfortunately the branches were occupied by leeches, plenty of them! Within a short time, the fitness levels of the two 'civvies' in the group became apparent. Don looked every bit the unfit 45-year-old middle-aged man that he was, and Herbert looked every bit the unfit 19-year-old youth that he was. I was giving both of them a hard time, quite deliberately acting as the hard Army man that would drive them on. But I noticed that my Army boys took them both under their wings and, to their credit, they were responding well. Don grumbled incessantly and Herbert gritted his teeth but neither of them suggested giving up.

Step by step we lifted ourselves out of the valley, onto the open plains. The wet button grass gently scratched our legs as we stepped around and over it, until it was replaced by stumpy alpine vegetation and eventually, quartzite. The quartzite spurs led up and up, revealing false crest after false crest; each crest opening up ever more spectacular views to the south and west coasts, and introducing us to the highland tarns. These tarns were deep lakes, ripped out by glaciers and now containing sparkling, dark blue water. We soon donned our Gore-tex jackets to keep the icy wind at bay but this caused the sweat to accumulate and so it was 'coats on' and 'coats off' most of the way up. We ate lunch in belting, biting rain, wondering if this was the norm in the mountains of the south west. Lake Tahune Hut, situated at the base of the final crest of the Cap, was our aim for the day and soon we were ensconced inside, chatting with some of the bushwalkers who had slogged for four days to be there, via the Loddon River Plains route. They told us that the dead pines we could see along the edge of Lake Tahune were killed in the big bushfires of the 1960's. They were pencil pines which take many hundreds of years to grow to any significant size. We were all gaining a sense of respect for the Tassie wilderness. This is how I recorded some of the moments:

> Don keeps asking me: 'How many more crests?'... and the answer
> I give is always the same: 'One more!' Upon arrival, the relief is
> enormous. We all have blisters. We all hit the sack. This is the
> hardest climb I have ever done. Herb has recovered his humour
> and is enjoying the place, taking photos. The blokes did well. Ian
> fell asleep on arrival and has remained that way since. Vic Shields
> sang all the way up—'The Sound of Music' would you believe?
> What a man he is, always cheerful, even when he is exhausted.
>
> Journal Entry, Ben Lans, Ex Olegas Truchanas, 28 Nov 81

Ian Biggs ended up sleeping for 17 hours straight! He missed the inspection visit by an enormous possum that stomped around the hut demanding a feed. Clearly he was the one in authority, probably appointed caretaker by National Parks!

Another journal entry:

> This morning there are swirling clouds around the mountain, ruining any plans of dawn photography. A cry of 'It's dawn!' by Don, was answered with... 'It has been so for two hours!' by one of the bushwalkers, whereupon all the gunners went back to sleep.

Journal Entry, Ben Lans, Ex Olegas Truchanas, 28 Nov 81

I had arranged to meet up with the lead Army helicopter the next morning because it was the first time that we had managed to communicate with our base at Strathgordon, some 55 km to the southeast of the Cap, since leaving Hobart. River groups were to communicate with each other and the base station daily, but this was always a questionable plan, considering the depth of the Franklin gorges, but from the top of Frenchmans Cap comms were excellent.

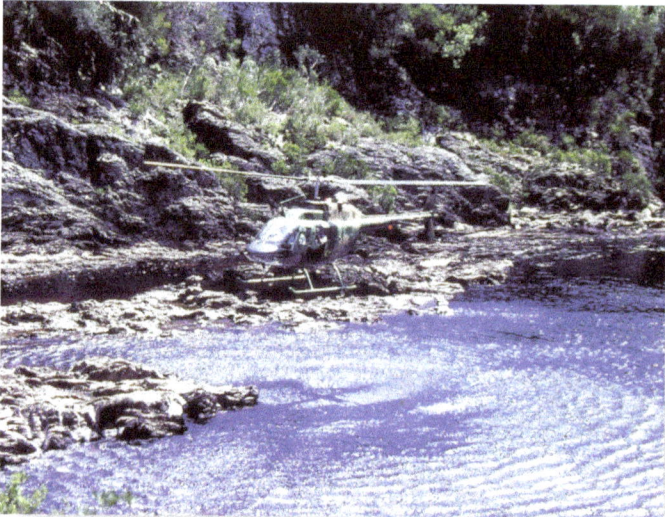

The helicopter dropping Group One back at the Irenabyss.
Photo Herbert Van Daalen

Despite the ribbing I was sure to get when the other groups found out what we did, I had no qualms in allowing my group to use the helicopter to do the last 'climb' to the peak, as both Herbert and Don were just about spent. After enjoying the breathtaking view, we were lifted back down to the Irenabyss. As predicted the other groups, who did not have the use of the helicopters, never let us live it down. We had made a large sign depicting '108' in the snow, a token gesture denoting that Army gunners had been there, and Herb took some photos as we took off. Then, we saluted the Cap and descended to the Irenabyss, where we bumped into Ian Johnson's group and shared the night with them. Mal Booth's group, which was second in the sequence, had already departed for the Cap. My entry:

> Back at the Irenabyss again, and Johnno's group awaits us. Heaps of stories and comparisons and another campfire. At the last rapid, Johnno ripped his raft open along its entire length and it was replaced tonight (by helicopter). Ian Biggs is providing great entertainment with his guitar. Vic is slightly pickled and is trying to sing Wizard of Oz this time. Sitting around this fire is a moment to remember. The feeling of companionship cannot be described... the fire is warm and a few rums makes it even warmer. To top this scene off, Johnno is standing by the fire with a handful of wet toilet paper waiting for it to dry so he can go to the toilet!

Journal Entry, Ben Lans, Ex Olegas Truchanas, 29 Nov 81

The warm inner glow of companionship evaporated quickly the next morning. Ian's group had rubbished the place as they left to go on their climb. I was angry at this lack of discipline. I trust people at face value. It's both a strength and a weakness of mine. I'm generally relaxed and don't anger easily, but the partially unwrapped rations and half packed stores scattered about made me cranky. My CO once wrote of me in his annual report... 'Major Lans' casual approach belies a professional attitude'. Well that morning my casual approach and professional attitude were both missing. I was unhappy with Johnno and I left a note to that effect. Then downriver a bit, I was sorry. I had ignored two golden rules. Firstly, always sleep on it and secondly, never write something down when you are angry. The anger, once written and delivered, is irreversible. When I saw Ian again he said that he had ripped up the note:

'Not in keeping with the Ben Lans I know!' he had said to himself. 'Water off a duck's back mate!' he continued.

We shook hands. I was not to know that dramatic, life-threatening events would occur with his group before we were to have that opportunity to shake hands again.

The next day, we accomplished Fincham's Crossing without too much misfortune. Vic cut his hand quite badly, but he remained upbeat and happy and I didn't think it would interfere with his paddling too much. It clearly didn't interfere with his ability sing! We were settling into the routine of the river and learning how to scout the rapids, read the water and the currents, and how to set ourselves up for a rescue in the more difficult rapids. That night Don cooked a damper which he offered to Ian, who described it's consumption as the equivalent of chewing a tube of wetsuit glue. A little harsh perhaps, as Ian himself had produced some memorable concoctions from time to time!

The men were beginning to show their personalities: Mark the medic, the physical one who assaults obstacles; Vic the organiser and the happy one, although we could have done with fewer renditions from musicals; Alan the planner, quiet and efficient one; Ian the joker and the carefree one, unstoppable, who would always do his utmost; Don the observer, who so accurately described people, places and events, acutely aware of everything around him; and Herb, the recorder, the maker of images, who was maturing by the day and never complained about the significant extra camera weight he carried. Cooking and eating was settling into a pattern of porridge in the morning, some biccies and a dry snack for lunch, and then something more substantial for dinner, like fried rice or pancakes, and always damper, which we took turns to cook. Vic was fast developing into the best cook.

It was that night that I wrote a letter to Lo. I wondered what she and the kids would be doing. They were in Sydney with Lo's parents, getting ready for Christmas. I was hoping to post the letter with one of the helicopter pilots.

Here is part of what I wrote:

I write by the light of a fire on the banks of the Franklin, one day past the Irenabyss, and two days from the Great Ravine. Frenchmans Cap is to our east, although we cannot see it. It is 9.00pm and still the twilight lingers. The noise of the river is continuous, as it has been for the past three days. This river has a strength and a character I cannot convey to you, either in a letter or in a thousand photographs. It is serene and beautiful, black and foreboding, frightening and peaceful... all at the same time. Here we are camped some 8m above the river surface, yet we look up and see driftwood 20m above our heads...

Ben Lans—Letter extract from the river to be posted on the next helicopter

The next day, we paddled to the Brook of Inveraestra, only about 14km downstream but right at the head of the Great Ravine. The weather was becoming consistently darker and darker, but no rain fell, as yet. This camp site was the one where I had to make up my mind: 'go or no go', because

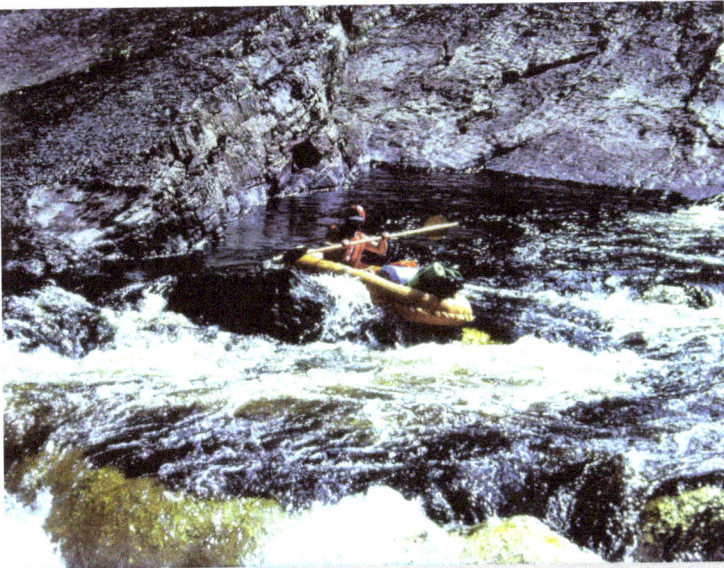

Bombardier Ian Biggs entering a set of rapids as the Great Ravine approaches.
Photo Herbert Van Daalen

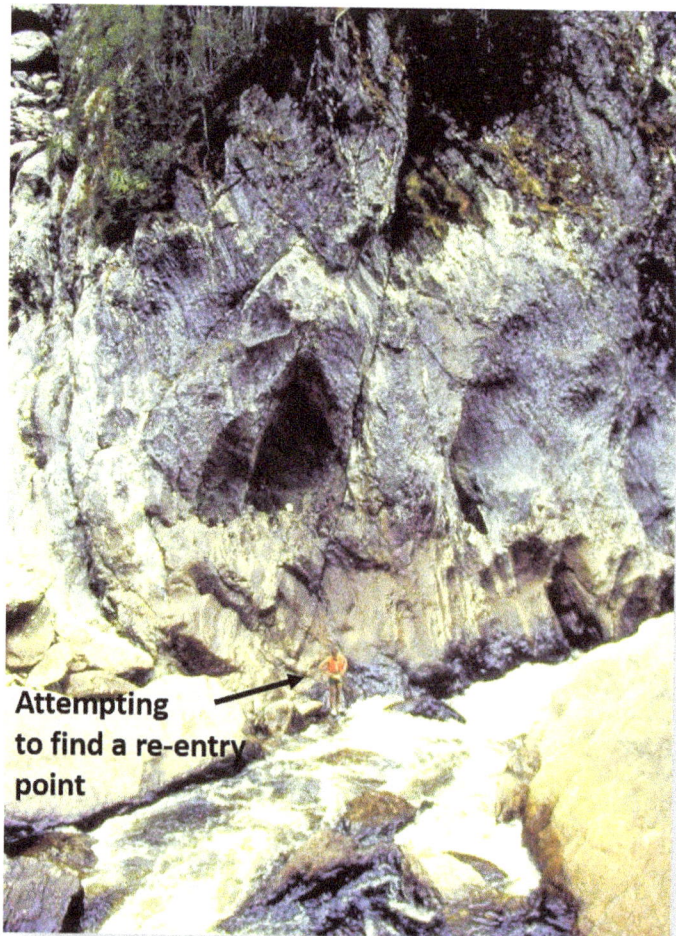

Checking the state of the turbulence at the end of the Thunderush Rapid. This photo illustrates the depth of the ravine and the manner in which the water is constricted, even at low water flows.
Photo Herbert Van Daalen

there were limited spots along the Great Ravine where a group can camp safely. Completing the Great Ravine in one day was not possible with these funny little rafts. So far, most of the campsites were quite cramped, allowing just enough room for four or, at most, five tents, with some of the tents invariably perched on tree roots or on a slope. This would make things difficult if we were to be caught by a group behind us.

Portage often involved carrying the rafts and equipment through steep ravines.
Photo Herbert Van Daalen

I decided it was go! On Wednesday the 2nd of December, we entered the Great Ravine. The walls here reach up to a height of 350m above the surface of the river, and the gap is narrow.

Throughout the ravine there are a number of major rapids mostly created by boulders, the size of houses, partially blocking the river. The numerous smaller rapids create fun 'chutes' and 'drops' to go over, but the four biggest ones, which are named The Churn, Coruscades, Thunderush and Cauldron, are all potential killers. They present up to Grade Six rapids, which is the highest (most dangerous) grade. Their degree of difficulty can change dramatically between low and high water flows, although the volume of water is not always the main factor that determines the risk associated with running a rapid.

Later, some of these rapids proved to be negotiable by skilled kayakers, but soldiers aboard little Taiwanese rafts do not make competent rapid riders. I had to limit my options and ensure that any risk taken was calculated to provide the greatest chance of survival. I had instructed the other two group leaders that these four rapids, apart from a portion of the Coruscades, were not to be attempted. There was no discussion on the subject. They were to be portaged. The day is best described in my journal entry:

Well we have come halfway down the Ravine, about as far as I expected to go by my schedule. Again, the weather has been overcast, although no wind. The dark sky makes the Ravine seem even more frightening than it is and makes photography somewhat less spectacular, or easy.

We are in a campsite some 10m straight above the river, with the sound of the Cauldron loud in our ears. This site is very rough, with many tree roots and small rocks. No room for tents tonight. The rafts are still tied on the water below. We hauled everything else up by rope. This site is near the portage track cut by Dean and Hawkins and party in 1958. I walked it in search of a better campsite but there isn't one. The track is very steep and rises some 100m through dense rainforest before paralleling the river and plunging back down. What men they were, these early pioneers! Even though the campsite is small and uncomfortable, the scenery is splendid. The close group atmosphere mingled with the presence of the river is never to be forgotten and we are warm and dry, with a hot feed inside us. It feels good.

But we're tired, dead tired. It was a hard push today. So much happened it's difficult to know where to start. Vic lost his paddle twice, once in a very precarious position at the end of the Thunderush, when he was about to be launched into an enormous boiling stopper. Herb leapt out in a spectacular fashion to get the paddle. We saw two snakes in the river, some weird 'sucker' eels, and a dead possum. Also a rat... or something. It shares the campsite with us. Its hole is inches from Herb's head. We have good communications with Propsting tonight, which is where Mal's group is. He called to say that they were well back on the river, some six or seven kilometres from the spot that we reached last night. They'll certainly not get far into the Ravine tomorrow. Still no word from Johnno's group.

We entered the Ravine with apprehension this morning. The first rapid of major size was The Churn. We portaged it on the left-hand side. Very steep track that took about an hour to negotiate with all the rafts and equipment. I was careless on the way down and my drum completed the last 15 metres or so by itself. The bottom, luckily not the lid, was crushed and had a hole in it, however the magic '100 miles an hour' tape did the trick and my drum still functions. Without a drum, life would be a misery. We portaged again just after The Churn, then allowed the rafts to make their own way down the Coruscades guided by ropes. Lunch was had in a very precarious position next to a six-foot waterfall.

Then there was Thunderush. A more powerful rapid I have never seen, even though this is low water. The ravine here is narrow and wild, with huge boulders strewn about almost carelessly. We climbed up the right-hand cliff first, and studied the frightening stoppers and awesome power of the water. I then climbed back, crossed over and studied the left bank for a portage. I settled at last for a left portage. It was a most difficult decision, as the start point on the left-hand side was quite dangerous, right next to a huge stopper with still a long part of the rapid to go. The alternative was a climb up the right-hand bank to a height of some 100m, which would take at least a half day. With my heart in my mouth I watched them all go, safely. Then Vic lost his paddle, once again to be recovered by Herb. Unexpectedly it surfaced right next to him about 50m downstream. Then it was my turn. I'd waited, as the last person would have no-one to help position the raft for the launch.

So here we are in the middle of this imposing place. How could anyone ever put this underwater by damming it? We're amongst the few very lucky ones to have seen it. Tomorrow we should reach Propsting where our communications party will meet us for a welcome 'hello' and a resupply of such things as muesli, chocolate and sugar. As I sit here, I watch 'double sucker-mouthed' eels swim past below us. Clearly a very busy stretch of the river.

Journal Entry, Ben Lans, Ex Olegas Truchanas, 2 Dec 81

That night I sat up late by the fire, listening to the river and chatting with Don and Mark. Well after midnight, we saw the lights of the Southern Aurora making strange twisting shapes of light in the sky to the south. For a while we didn't know what we were looking at. Such beauty! It occurred to me that the blokes had all been commenting on the magnificence of this place. Perhaps it was because all my soldiers knew how passionate I was about preserving wilderness and they didn't want to offend or upset me, but it seemed to me that most of them shared my views that the Franklin River and its surrounds must be saved. We were bunch of Army gunners trained for war, but hey... even gunners can appreciate raw nature in a wild environment; just because it is there! A place lightly trodden. A place both simple and complex in its beauty. A place of adventure! A place worth fighting for in all meanings of the word.

We camped without tents that night: there was just no room. But it was here the rains started. Most of us had prepared for it. It's part of our training... the well-known Army saying: 'Any fool can be uncomfortable' comes to mind. Unfortunately, Don, who'd actually been a part-time soldier once upon a time, hadn't remembered that lesson. We listened to him bellow about in the middle of the night, grumbling about no-one caring whilst trying to unpack his tent fly to wrap around him. I considered helping him, not because I felt sorry, but I worried about him stepping over the edge and plunging into the river, to be swept into the Coruscades rapids just below us. I didn't have to worry for long. Good ol' Vic got up and helped him. Don finally hunkered down, curling up like a cranky, wet dog. You could always rely on Vic.

19 THE MIDDLE FRANKLIN

The end of the Great Ravine was not far away, but first we had to tackle The Cauldron, the last Great Ravine rapid, which would lead us into the area that Bob Brown named 'Deliverance Reach'. By his description, I expected the ravine to widen at this point, and the river to take on a more 'friendly' persona. There were still many rapids to come, including some extremely dangerous ones that would need careful scouting. Once we passed Rafter's Basin, a large pool at the confluence of the Andrew and Franklin rivers, we hoped to meet with our support crew in Propsting Gorge. The Hydro had constructed a sway bridge there, accessible via the Mt McCall track which is connected to the sway bridge by a rail cableway, rising at about 20 degrees from vertical and approximately 150 metres in height. It comprises a series of sleepers which, with care, can be used as a super-long step ladder. Good exercise for the 'loggies', who had been sitting on their backsides eating fresh rations and doing the occasional radio checks, or so the fellas all reckoned. It's the only spot along the entire river where a vehicle can approach and was built by the Hydro to provide access to their proposed dam site. If they were to build that dam, there will probably be a small town up there, with workers and machinery enough to build a concrete monstrosity. The thought of this wilderness being invaded by the destructive forces of the Tasmanian Hydro was obscene.

We packed in silence with a sense of trepidation and the roar of the river in our ears. We were about to be tested again. The Cauldron comprises a series of house-sized boulders which create turbulent holes and stoppers, where the current is forced under and around obstacles, creating dangerous hydraulics that can capture and turn a raft, spilling the occupants into the fast-flowing aerated water and potentially trapping them under or against rock ledges.

I decided the water level was sufficiently low to enable us to rope our rafts down whilst rock-hopping beside them. This meant hanging on to a tie-line at the back of the raft, leaving all the gear inside and letting it find its own way down. Although that sounds easy, it involved risk. The rocks we 'hopped' were nearly as big as houses. One slip could send any one of us into the raging torrent to an almost certain death.

One by one we made it safely. It silenced us all—being only a wrong step away from drowning in boiling aeriated water.

Soon we were in Rafter's Basin where, in the company of two large tiger snakes, we had lunch. Fortunately, the tiger snakes were warming themselves on the rocks and showed little interest in us. After lunch we drifted on gentle waters and marvelled at the swirls created by the foam being washed out of the Great Ravine. This was a time to relax and feel privileged to be in a place that very few had ever seen. Despite the serenity of the moment, there was a persistent, nagging thought lurking in the background. How was Ian's group going? They were last and should only have been one day behind us, but we had heard nothing from them—no radio communications at all.

Not long after we entered Propsting Gorge, we were reminded that the river was not about to give up its hold over us. It continued to present an air of lurking danger, even as it displayed its beauty. This was not a time to relax, as my journal entry of that day explains:

> A small mishap occurred after we left the Ravine. Ian Biggs had become a bit blasé about rapids, drifted into a seemingly harmless one, and was instantly upside down under a log, raft and all. Herb stopped to help him but joined him instead. Two rafts jammed.

Alan Ward, demonstrating how to stay in his rubber ducky in the Propsting Gorge rapids.
Photo Herbert Van Daalen

Two bodies under water. A comic scene initially but urgent action was required. Mark and Alan pulled and tugged at the rafts, which would not move. At last they did! Fortunately, both Ian and Herb surfaced quickly. Herb's paddle was broken. Mark and Alan are proving to be the strong men of the group. Quietly capable and quick-thinking.

Journal Entry, Ben Lans, Ex Olegas Truchanas, 3 Dec 81

We made camp under the incongruous sway bridge, a blatant invasion into this wild place, and settled in to await the expected re-supply. Within moments we heard that most famous of Aussie cries:

'Coooo-eeee!'

Not very military maybe, but very appropriate for this occasion. And there he was: Sergeant Harry Pregnell! He didn't let us down, having crystal-balled the precise time that we would be here, based on only a few intermittent radio transmissions. Harry was well met, handing out fresh food and chocolate bars, cracking jokes and catching up on all the stories. He planned to meet all three groups here. The meeting was executed with clockwork precision, like just about everything else that Harry did. Mister Reliable himself. We gorged on the extra goodies then waved the support boys goodbye as they embarked on their long ladder climb back to the top.

My aim was to make the next camp at the caves alongside Newlands Cascades, but there were a few major obstacles to overcome first. Bob Brown had warned of the dangers of a rapid he named the 'Pig Trough' where a chute of fast water leads into a gate between two boulders, followed by a sharp drop, creating a treacherous stopper with a boil of water so large that it would turn a raft back on itself. Once caught in a stopper of this size, the only way out for a rafter is to dive deep and attempt to find water exiting along the bottom. This assumes that the rafter is capable of making such a rational decision in the panic of the moment, and that there are no traps, such as submerged rocks or trees.

Just beyond the Pig Trough is Rock Island Bend, which marks the beginning of the very last rapid in Propsting Gorge. This rapid, Newlands

Harry Pregnell and the ever reliable logistic support group climbing over a hundred and fifty metres down, using the near vertical railway line originally built to support dam construction, but never used.

Photo Herbert Van Daalen

Cascades, was named after one of Dean and Hawkins' group where, in 1958, on their third attempt, they realised that they had successfully negotiated the worst of the Franklin. We portaged around the Pig Trough and, one at a time, enjoyed the many chutes and drops and holes and swirls of the cascading water, which continued for hundreds of metres, until the rapid spewed us out at the end, onto the flat waters of the lower Franklin. There we dragged our rafts ashore and hauled our gear back up the rock ledge running along the cascades to find the caves. These are formed by a delightful, natural rock overhang which provides cover for at least a dozen or so sleeping bags, with

The camp in Propsting Gorge, where the support party met us, on one of the few sandy beaches in the middle Franklin.
Photo Herbert Van Daalen

ample room for drying gear on the rocks and space for cooking an evening meal in the sun. It was time to dry out and soak up some sun!

My journal entry:

> Propsting Gorge is a stunningly beautiful and peaceful place, not as deep as the Great Ravine but less intimidating. How the sun makes the moss stand out, the rocks shine and the water glisten! I sit here on a rock ledge above the river and just near my feet are bush orchids with purple flowers, and some with white flowers. We made camp at about 5 pm today—it has been a hard day, but an exciting one, having portaged only one rapid, called the Pig Trough, which has a nasty 'killer' stopper at its base.
>
> Many of the rapids consisted of a series of drops around one to two metres high, usually in quick succession, with a few stoppers thrown in for good measure. Every rapid caused at least a couple of us to come to grief. I did a spectacular backward somersault out of my raft today, and rode the rest of a rapid, known as 'Ol'

Three Tiers', upside down, bumping my head against submerged rocks. I was thankful for the helmet. My paddle broke again, but it was able to be fixed with tape and tent pegs. We got some great photos. I kept sending Herb ahead to climb the gorge walls to set himself up for photos, which he did, and he took some great shots alright, but then he would have to come down at the end, always last. We made Herb work for a living.

Don's rapid-riding style remains the same—the 'stiff dolly' look. I did admire his courage as he was pretty scared of the rapids, but he trusted my every word and plunged in with abandon, not wanting to lose the respect of the gunners. Mark and Alan were their efficient selves and Biggsy seemed to think he was invincible. Vic was the funniest of all today (at Ol' Three Tiers), when he launched from his raft, flicked up into the air by the whipping effect of the racing, lumpy water, and seemed to sail a metre above everything. He then crash-landed back into the raft, skinny legs and arms flailing, and came up with a smile, albeit a slightly nervous one.

The stoppers had a habit of pulling us up to a dead halt, then dragging us back allowing the fall of water to fill the raft and turning it into something akin to an elephant in water. Impossible to steer. The adrenalin ran free today. These were the most 'fun' rapids thus far.

The rapid called the Pig Trough is supreme, though it has a most unfortunate name. A creek drops in from the north in two stages, about 40m in all, the resultant mist covers every rock and mossy outcrop with glistening drops, making the going very slippery and precarious. Downstream, less than 50m from the junction, is Rock Island Bend, a spectacularly huge boulder creating an obstacle in the middle of the river, causing the foam trails from the Pig Trough's stopper to bend in graceful arcs and wend their way around the bend, making all sorts of patterns on the water's surface. Just around this bend is a lengthy set of rapids known as Newlands Cascades. These are fast and clean rapids and quite safe and we negotiated them with ease.

It is almost dark now. We will sleep well tonight. As a group, we are beginning to feel we understand this river and we have become a part of it. We have no right to say this though. It has been here millions of years and we pass in but a fraction of a heartbeat of its life. And yet we as humans have the capacity to destroy all this. I feel deeply sad.

Journal Entry, Ben Lans, Ex Olegas Truchanas, 4 Dec 81

That night I slept so well that it was a shock to hear someone calling my name. It was only just light!

'Boss... wake up. We've had a drama! Herb nearly drowned, and he's ruined the cameras!' It was Alan.

'What? Waddayamean?' I struggled to make sense of the bad news that was trying to penetrate my sleepy head!

I looked away from Alan's face and focussed on Herb standing behind him and immediately knew something bad had happened. Herb's demeanour said it all, standing there all bedraggled, dripping wet, holding a bunch of equally wet cameras in his hand.

'Oh shit!' I said. 'Herb you *have* actually drowned the cameras!' and as my brain focussed, I added belatedly... 'and you nearly drowned yourself? What happened?'

He drowned the cameras alright, ruined them: all three Minolta SLRs plus the lenses. Catastrophe! Herb had a rush of blood to the head before first light and decided to mount every lens he had on each of the three cameras and carry the lot with him, to take some early morning shots of Rock Island Bend, all on his own! He said he was climbing around the base of the rock when got a bit cocky, tried to take one step too many, and slipped into the water, with the whole caboodle around his neck!

I was angry with him! How could he do such a thing? How could he take such a risk with his life without telling anyone what he was doing? He could easily have drowned amongst the steep, slippery moss-covered rocks and fast-flowing water smashing against the rock. And now no cameras! My soldiers would not have done such a thing. And again I let him cop my wrath! But... then I remembered that Herb was not actually one of my soldiers. He was not trained to be a team player, he was just a kid. Suddenly his youth and lack of experience jumped into my reasoning and I saw that he was in mild shock. I needed to go easy on him.

In the end he copped it sweet and accepted the dressing down I gave him but I'm sure he felt justifiably put out that I seemed to care more about the cameras than him! That was not the case, but I didn't tell him that at first! Later I assured him, somewhat jokingly, that I really did care more about him than the cameras, but only just! It was a character building moment for him and for me. Herb gained several years of life experience on the spot and I probably gained a few grey hairs. I realised that I had been expecting way too much from my young nephew! I looked at Herb with new respect after that incident. He was plucky!

By some amazing stroke of luck, we managed to raise Harry on the radio and in no time at all, out came the helicopter with none other than Tracker on board, to pick up the drowned cameras. In only the way that Tracker can do these things, he casually hung out of the chopper door whilst it had one skid on a large flat rock in the middle of the river, and the other hovering above the raging torrent. The space available near the cave overhang was insufficient to accommodate the chopper blades, so the only way to pass anything to the chopper was via a raft, with one person paddling and another handing the bag of cameras to a person on the chopper as the raft sped by. This was no easy task as the rock upon which the helicopter partially settled was in the middle of the Newlands Cascades! We watched in awe as the pilot performed a miraculous balancing act, allowing Tracker to reach out using the air borne skid as support. and grab the bag of cameras being offered by Mark, who was standing up in a fast moving raft paddled by Alan, whilst holding the bag somewhere above his head, trying not to fall overboard.

'Bloody hell... if only we had the cameras to record this. No one will believe it!' said Biggsy.

'Perhaps that's a good thing,' I mumbled.

The news that came back via Harry was all bad, at first. The cameras were unrepairable. Herb was inconsolable, so I took the opportunity to give him a hard time by telling him I was changing my mind again as to what was more important, the cameras or him. However, Harry came to the fore again and, incredibly, managed to contact the Minolta product manager on a Sunday afternoon to explain the situation. This was Derek Plante, my brother-in-law, who contacted his General Manager, who happened to be having a barbeque at his house in Melbourne and who promptly authorised replacement cameras and lenses. Even more remarkable was the fact that Harry managed to arrange for the replacements to be made available from a friendly camera store in Hobart! The replacement cameras were to be delivered a few days later. Herb was saved: we kept both Herb *and* the cameras. Although we made light of the whole thing in that typical humorous logic that soldiers apply to everything, the fact that Herb may well have drowned before any of us knew where he was or what he was doing, would stay with me!

20 THE LOWER FRANKLIN

Early this morning it started to rain... it absolutely poured, accompanied by a cold wind that cut us to the bone. We packed reluctantly thinking we might find a cave downstream. We passed the entrance of the Jane River which Ian checked out, but no campsites, and in any case, every bit of it was soggy and wet and not conducive to putting up a tent.

Journal Entry, Ben Lans, Ex Olegas Truchanas, 5 Dec 81

We left the caves at Newlands and entered the world of the lower Franklin. The river was now much wider and slower, making the paddling harder, the going tougher. Gone were the steep sides of the various ravines we had been paddling through, replaced by sandy beaches, and rapids replaced by riffles of water. Since it had begun to rain consistently, my mind turned again to the other two groups. Where were they? Would they be able to clear the Great Ravine in time? It's not unusual for the water level to increase overnight by 30m in the Ravine. It only has to rain an inch or two somewhere in the highlands, and the steep, narrow nature of the ravine causes the water level to rise rapidly. I worried that one of the groups would get caught.

'Hey Ben, this rain is giving me the shits,' a voice from the back belonging to Biggsy.

'Okay, I think there is a hut here somewhere' I said.

'You're having me on!' came back the reply.

But I wasn't. I'd kept this one up my sleeve. Marked on Bob Brown's hand-drawn maps was a hut at the end of a track that leads across from the Gordon River to the Franklin. Alan spotted the track leading away from the river and there it was, the rudimentary Hydro hut with a fireplace and a roof! In the midst of all this rain, we located the only habitable hut on the entire river, built by the Hydro for use by their technicians when they came to measure the river height. There was a tinge of irony in all this. Here we were not wanting the Hydro to dam the river, but at the same time we were overjoyed with the warmth and comfort that the hut provided on these persistently rainy and cold days.

We hunkered down and prepared to wait out the rain. I was still none the wiser about the other groups until the next day when Mal Booth's group arrived, sodden and cold. At least I now knew where he was. They were hopeful of staying with us in the hut but 13 people in a hut designed for two wasn't possible. Reluctantly Mal took his boys away again and they camped on a small island called Rat Island, right there in the river, opposite our warm, cosy hut.

By now the river was rising fast, threatening the safety of our rafts. They had been tied to the bank near the track to the hut and were now straining at their moorings, half under water. If the lines had broken, they would have been swept downstream. Ian Biggs was the one to remember that the rafts may need rescuing. Two rafts that had already broken free and could be seen caught in the rushes and trees on the bank, a hundred or so metres downstream. Mark ferried him into the maelstrom of swift running, swirling water, to retrieve them. In a series of determined paddling efforts and heroic swims in the freezing water, Mark and Ian managed to cut them loose, secure them, then work their way back by dragging them up the riverbank against the current. That was close! We nearly lost those rafts!

'Well done fellas. I'd give you a slap on the back for your efforts but you're all wet and I'm dry, so bugger off and stop interrupting our chess game!' I said, tongue in cheek, very appreciative of what the blokes had just achieved.

Like the Herbie incident, the gunners made light of that event as well. It's a beautiful thing to watch a bunch of blokes work tight... together, effortless in everything they do, say and think, to achieve an unspoken, common goal.

For three days we stayed in that hut. At first it was good to feel dry and warm, but we were here for adventure and we started to feel guilty that we should be sheltering like this whilst the others were cold and wet on the river somewhere. Indeed, I still had no real idea of where Ian's group was. Were they safe?

We had so much time we had carved a chess set to play our games of chess! I also took the boys on a walk along the track that heads back to the Gordon, the ones used by the Hydro people to access this hut, just to fill in half a day. This was an uneventful walk except for the fact that it rained and hailed, and the walk had absolutely no purpose to it, other than to make everybody snap out of the comfort zone they had become used to. This is how Don Johnson described the events surrounding the stay in the Hydro hut:

6 Dec. After (passing) the Jane (junction) it began to drizzle. Heavy forests choked the banks and we began to look for a campsite. Ben and Ian recce'd one. I had already begun to unpack my raft when they changed their minds. Downstream a large scar on a cliff. Apparently, there is a hut. I sat wet and utterly despondent while they went and checked it out. The rain has begun to sheet down. Phrases like 'open fireplace' and 'bed-spaces' floated tantalisingly down from above. We decided to stop. I hauled my drum up a very steep, slippery 10m slope and Ian Biggs came to help. I was too cold and too exasperated even to curse as the bloody thing cartwheeled back down. Ian handed it back up to me, twice as heavy and I emptied the water from my totally saturated gear. They had a roaring fire set already... I sallied forth in my thermal underwear to gather wood...by 1500 hours we were warm and comfortable and now the card game in the corner has sunk to a low murmur. Vic is entertaining everybody

with his version of 'chicken little' and Ben is drowsily writing up his notes. I hope that what I have written is sufficient to jog my memory: I'm generally too bloody tired to try for descriptive prose each day.

7 Dec. Asleep to the comforting sound of rain. Awakened by Vic scrambling to call the helicopter overhead at 0900. Ben callously orders Biggs from his bed to the raft and Ian stoically obeyed. Harry Pregnell materialises smiling with two new Minoltas and lenses... the rain keeps coming in sheets. Ben decided we should all walk to the Gordon; we took off at 11.30. We found another old hut some 100m along. The rainforest was wet and the track, because it was open to the sky, was covered in moss. There were alternating showers which, before we left the forest, turned to hail and the track as it opens up into button grass, allows freezing winds and hail into our faces. We walked into the teeth of it. Icy pea-sized lumps. I have rarely been colder. The track down followed a spur line. We halted at 1500 and reversed. Rain and hail on the return journey, it lay across the moss in drifts. I was okay on the flats but the hills really ironed me out. Vic lingered behind to make sure that I made it. I staggered back. It is still raining and the river is rising.

8 Dec. Ben has made the decision that we will abort and run for Butler Island to RV with the Denison Star... the trip will end with a bang. The Franklin was overkind to us in the upper reaches but she is as slippery and as muscly as a python outside tonight. It is a matter of pride to the soldiers that they face a dangerous finale. I am not unhappy to be finishing the trip on a high key... the Great Ravine is burned indelibly into my memory. The rest of the scenery was splendid but only that stretch matched my deepest foreboding.

Journal Entry, Don Johnson, 6-8 Dec 81

On the third morning of consistent rain the weather cleared and both Mal and I packed up our groups and headed off, Mal first.

This was now a completely different proposition. The Franklin was up and roaring, flooding the surrounding tree-covered banks. As soon as we hit the water we found ourselves speeding along at a pace that can't be matched by any paddling, on any type of craft. We were riding an ocean of water pouring straight out of the Tasmanian highlands; no time to relax, impossible to stop! There's no point looking for sandy banks because we were going too fast, and anyway, there were no sandy banks visible: they were underwater, submerged by a violent, angry river. We barely needed to paddle, so we reflected on our surroundings.

The rainforest in the lower Franklin was different to what we had been experiencing: no longer the myrtles and the sassafras and the soft gentle drooping branches of the Huon pines. Instead, gigantic eucalypts dominated the banks and the land beyond, as far as it was possible to make out. They created a dense canopy, shielding the darkness of the forest floor below, which received only filtered sunlight at best, and appeared to be permanently soggy, with almost impenetrable undergrowth. Numerous tall, thin trees had tried to reach up amongst the dominant giant eucalypts to find the sun, but failed. Instead, they lay on their side, apparently exhausted by their efforts. These thin trees then shot branches straight up, once again looking for the light and failing, falling again. This pattern was repeated until the undergrowth appeared to be one mass of interlocking trees and branches, forming a formidable barrier to anyone trying to penetrate this wild place. This is what Tasmanians call 'horizontal scrub.' I wondered how the escaping convicts managed to get anywhere at all in this place, let alone find their way right across the island to the east coast.

Mankind had in fact existed here for millennia, because it was in this area that Kevin Kiernan, a geomorphology student, found a cave in 1977, only a few years before our trip down the river. The cave, subsequently called

'Kutikina Cave', revealed important archaeological deposits that proved the most southerly human occupation on earth during the last ice-age, which ended about 15,000 years ago. It also revealed evidence of wallaby hunting at a time when the landscape was an open tundra. I had been given the approximate location of the cave by Tasmanian Parks and Wildlife and had arranged for an archaeologist, along with an ABC news film crew, to meet us in the vicinity of the cave using our helicopters. The cave location was not public knowledge as yet, and the archaeologist was prepared to do a short news item of soldiers being shown around the cave. It was an invaluable opportunity for him to use our helicopters to visit a place that he was clearly passionate about. Not only would we be educated about the cave and its history, it would also have presented another opportunity to focus national and international attention on the Franklin River: on what might be lost if the river was dammed. The publicity would also have been good for the Army and my sponsors!

But it was not to be. The river was in flood.

For us, however, there was no time to contemplate anything other than the here and now. We were too busy just staying afloat and I knew we were not yet done with the rapids. There were two tricky rapids to go: 'Big Falls' and 'Little Falls'. What I didn't anticipate was that the river was so high that both were completely covered and unrecognisable. I was about to learn the lesson that high water changes everything. The dynamic hydraulics of water flowing fast at high volume is unpredictable, dangerous, and should be avoided.

Quite suddenly, as we were belting along, my senses were alerted. On top of the already frightening freight train sound of the river came another, even more extreme noise.

'Watch out boss!' came the screaming voice of Alan, as he stared ahead. 'Take a look at that wall of water!'

Everyone turned to look, then turned to me as if I could somehow explain what it was they were seeing and manufacture a plan in a matter of moments!

I could barely hear him but the urgency in his voice was enough. We were paddling in a reasonably tight group, covering perhaps 20 metres of river between us, all moving very fast, swept along by the current covering many metres in seconds. Ahead was a constriction in the river that caused the water to boil violently upwards.

I stood in my raft, balancing precariously for a moment, and saw enormous standing waves of water, known as 'haystacks' in rafter's language. Each one had a 'curler' at the very top, resembling the crest of a wave on a surf beach, that was aimed 'upstream', looking as if the wave might break upstream instead of downstream. These haystacks are unlike ocean waves, which move forward whilst the water remains relatively still, merely rising and falling in place. Haystacks stand in a fixed position while the water rushes through them. There they were, one after another, after another, apparently stationary, but exploding upwards! The first wave looked like it was three or four metres high, and each subsequent wave only fractionally lower. This must be Big Falls. However, instead of a waterfall with water dropping over an edge, the high river had swallowed the falls completely and they now consisted of standing waves, belching upwards! The waves, the foam, the deafening sound... disaster approaching with no time to think!

This was a clear case of, if the first wave doesn't get you, the next one will, or the third or the fourth! The whole scene was like a river exploding: waves, wind, noise, with us rocketing inexorably at this mayhem of nature ahead. Catastrophe was inevitable: 'What now, Team Leader?' ricocheted around my head. I had seconds to decide. The alternative was to head for the banks immediately, but the riverbank offered little to help us: thick vegetation and tall grasses would simply tangle the rafts and more than likely tip us out. I didn't fancy being swept into the maelstrom of mad, foaming water without a raft. Better to ride this lot on top of a raft than beside or under one.

All the men were back-paddling frantically waiting for a decision.

'Okay, let's go for it,' I yelled at the top of my voice, as I dropped into my paddling position, accompanied by much gesticulating and general signalling of commands, which I hoped would make everyone understand where and how to try and line up.

'I'll go first—then Ian!' He was right next to me and he could at least hear me.

Ian suggested we try to pull out at the end of the worst waves to perform any rescues that may be needed!

'Ian, good idea mate, we'll pull out, left and right ... at the end! Try and find an eddy! Anyone tips out we'll grab 'em!' I screamed at him. We were now just tens of metres from the point of no return.

'Good luck, mate, and don't *you* fall out!'

And that's all there was time for. Off we went.

The others didn't hear those instructions to Ian but realised what was going on and lined up as best as they could in the fast water. Heart in mouth and adrenaline pumping, Ian and I made it! It was surreal. Frantic paddling to reach the top of each crest, then smashing through the curl. The rafts, with packs and drums in the front, had just enough weight to pull us over the top and down the other side, before the frantic paddle resumed! All this water, all this noise, all this... ferocious nature going about its business! And here we were, mere humans, daring to be part of it, with our pathetically inadequate equipment. We looked like bits of modern-day flotsam at the mercy of a wild river that is all powerful. After cresting at least five or six waves, we managed to find eddies, one on each side, and there we sat, hearts thumping, lungs pumping.

One after the other, the men came through safely, until it was Mark's turn. He was last to come through and was unlucky to be hit by a strong gust of wind just as he crested the first, and biggest wave, and over he flipped! His raft continued, albeit upside down, but Mark himself was unseen. Ian and I swung into action, paddling straight into the centre of the current with a desperate urgency to stay in one spot. We waited, and waited, and then suddenly up he popped! Mark, with big eyes and mouth gulping for air, his arms barely moving and unable to stroke to keep himself afloat.

His life jacket now kept his head out of the water, but the shock and the effect of the extreme cold water had already started to shut down his muscles and render his arms almost useless. Ian grabbed him by the scruff of the neck as he swept past, but it took the two of us to haul him out of the water and into Ian's raft.

That cold water made short work of a very fit man very quickly. After only a minute or so in that river, Mark was incapable of assisting his own rescue. I was shocked that anyone could lose so much strength almost immediately but felt thankful he was safe.

We eventually managed to get him back into his own raft and we continued hurtling downstream. It was imperative to stop and get some hot brew into him. He appeared to be going down with hypothermia and he was the medic! I couldn't have that! I planned to stop at the junction of the Franklin and Gordon, on an island called Pyramid Island which normally rises several metres above the river surface, but it was nowhere to be seen.

'Clearing, over there I think, just through that lot,' said Alan, pointing to the bank near the junction.

Without being told, everyone headed over at once, glad for an excuse to stop and make a brew. It turned out that Mark had gashed his leg quite badly which needed dressing.

From here it was just a short paddle down the mighty Gordon to some Hydro huts that were marked on our maps, near Sir John Falls. It was here that the Hydro planned to build the other dam, the bigger one, and it was here a few years later that the protests would be held to successfully stop the dam's construction.

Nearby was Butler Island, a small but prominent island where the tourist ferry *Denison Star* came with its boatload of tourists every day. I'd made arrangements with the captain of the *Denison Star* to pick us up in the vicinity of Butler Island and deliver us to Strahan. As we pulled into Sir John Falls, we were met by Mal's group. They too had sped out of the river, hurtling along in the same manner as we had, but without the unfortunate dunking at

Big Falls. It occurred to me that we had seen nothing of Little Falls. Clearly, they had also been completely swallowed by the water, but without the creation of standing waves. The lower part of the Franklin would normally have taken us three days, we did it in about five hours, a testament to the speed of the river in flood!

The trip, but not the adventure, was nearly over. Here is how Don described the last moments on the river:

> As we approached the end I dropped a couple of hundred yards behind. Not last. A rope on a tree on the south bank and the sight of rafts landed. Mal's party, at a veritable chalet (the Hydro huts). Landed the gear, selected a sleeping spot and changed into dry clothes. A sense of sadness I suppose, at a fellowship diluted. One of those ephemeral unions which, among caravans one joins on the journey will remain a bright memory. The two groups laughed and exchanged stories, already selecting and codifying their memories. Mal and I exchanged views and Ben became more impersonal, the public leader, not the boss.
>
> Journal Entry, Don Johnson, 9 Dec 81

Now that two groups were safe, my mind turned again to the other group. We still had no contact and I knew that they were at the mercy of the rising waters in the Great Ravine. Where were they? Were they safe?

21 What Happened to Group Three?

As we landed at St John Falls and pulled our rafts out of the river, Mal approached me. He had already communicated with Harry by radio, and the news was not good. I'd been unaware of what was happening to Johnno's group. Mal said that they'd suffered major drama. The whole group had been caught in the Ravine as expected, but not in a safe place. They hadn't fared well. The information he had was scant and my head raced with stories of helicopter rescues, lost equipment, injuries, but thankfully no deaths. All the 'what ifs' started racing through my mind. Had I briefed Ian sufficiently? Was I confident he understood the dangers? Was my communications safety plan good enough? Should I have deployed more back-up people, more communicators?

I didn't have to wait long to hear more of the story, because we suddenly heard one of the helicopters was coming in. Somebody threw coloured smoke on the little landing pad at the end of the camp and the pilot honed in on it. One colour suddenly morphed into two, then three, and soon a kaleidoscope of colours wafted into the clean Tassie air! We'd been carrying the smoke canisters for just such an occasion and the gunnies were happy to be rid of it. Patiently we waited as the pilot shut her down and the rotors slowed... and then, out stepped Sergeant Peter Nolen, gingerly holding his ribs. The gunnies swarmed around Peter, but the pilot indicated he wanted to update me.

The Army Kiowa delivering Sergeant Peter Nolen to the river end point after he was plucked from the Great Ravine. The rest of the group was ferried back to Hobart.
Photo Herbert Van Daalen

First things first, he said. Everybody is okay. There were some minor injuries but nothing too serious. Peter Nolen, whom he had just delivered, had suspected broken or badly bruised ribs and he had brought him here for Mark, our Medic, to check him out. I glanced over and that was exactly what was happening. Mark was examining Peter's ribs to the accompaniment of most of the group, who were crowding around trying to find out what had happened to their mates. The pilot went on to tell me that two of Captain Johnno's group had been evacuated to the Strathgordon forward base with minor injuries including scratches and bruises, suffering from mild exposure, to be treated by the medic there. The rest of them, three in total, were still in the Great Ravine and would be rescued the next morning. There were not enough daylight hours left for helicopter operations in such challenging terrain and the pilot needed to get back to refuel, ready for the morning's flights. Everything was determined by the availability of fuel, which is why he flew Peter Nolan to my location for our medic to attend him, rather than the Strathgordon site. From here, he had only enough fuel for a straight run back to Hobart, without a passenger, so if Nolan needed hospitalisation, he would have to wait till tomorrow!

Moments later the chopper lifted off and as I watched it turn into a small dot, I recall taking a deep breath. Ian Biggs came up beside me and handed me a brew.

'Here mate,' he said. 'Here's some strong coffee, the way you like it. You look like you need it!'

'Jeez, Ian,' I said after one sip. 'This'd make your grandmother roll over in her grave!'.

'Yeah, Harry taught me!' he said with that typical Biggs smirk, and he continued: 'We're a team!'

'Both of you'se trying to poison me!' I said.

Who could have imagined it? When Army Headquarters told me the trip would not be approved without helicopter support for safety, I was scathing. Why was it a 'must' that we get Army helicopters? Over the top, I thought!

It was not impossible that such a rescue could also have been done with civilian helicopters, but potentially not with the same result and, in any case, a civilian helicopter would not have been patrolling the river like our military ones. Here is the story, as derived from the bits and pieces of information I later gathered.

Johnno's group, as the last of the three groups to negotiate the Great Ravine, was supposed to be 24 hours behind my group. After that meeting at the Irenabyss, they slipped to being about 48 hours behind, but I didn't know that because we had never managed to achieve communications with them. Was their radio working properly? We never found out because that radio was lost in the Great Ravine. When we were sheltering in our Hydro Hut and Mal's group were on Rat Island—and the rain belted down incessantly—Ian's group was trying to negotiate the Great Ravine.

The raw truth is he should never have attempted it. He broke the golden rule: when it is raining heavily, do not attempt the Great Ravine because the likelihood of being caught out is almost certain. There are very few safe spots within the Ravine to escape rising waters—you can't just stop, because the sheer walls prevent that. And in a rising river it's often unsafe to proceed, as the rapids become cauldrons of boiling, foaming, angry water. I agonised over whether or not I had briefed Johnno sufficiently about the dangers of proceeding in high water.

In any case they went ahead and, after they entered the Great Ravine and portaged the Churn, it became too late in the day to continue, so Johnno decided to look for a camp. Finding a safe haven for the night, already nearly impossible, was made more difficult by the fast, rising water and the looming darkness. In desperation he chose a log pile on the northern bank. It didn't occur to anyone in the group that the logs had been piled there when the river was in flood at other times, and that it was a vulnerable place. However, in their defence, even if they had understood the risks associated with that site, it was unlikely that they would've had any choice in the darkening light of a rainy day, in a ravine with sheer walls. The group clambered out onto the logs, but quickly realised that the river was following as they climbed higher and higher up the logs—they were unable to create a camp. Some tents were established however, and they settled in for the night, jammed together, unaware of the river's ability to rise 30 or 40m in a few hours.

In no time, they were awakened by the increasing roar of a river entrapped, a river rising, ready to drown them if they didn't move. This is the gist of what one of the gunners told me:

> 'The night was bloody pitch black and all we could hear was this bloody roar. We couldn't hear each other speak! Couldn't yell! No point! Didn't have a clue where the others were. Couldn't find my gear. I was shit scared!
>
> The boss was doin' a head count and tellin' us all to climb up the bloody log pile and stay with the person in front. He was really calm. Couldn't see a thing so it was all by touch. It was like staring into blackness. He'd done a recce just before it got dark and

reckoned that there was a safe route upwards... told us to leave our kit but make sure we take our wet-weather stuff and sleeping bags, just so that we could get dry. That freezing water was following us up, would you believe? We did as he said... no choice really, but it was bloody hard stumbling in the dark over these massive logs. They were piled almost straight up! We slipped a real lot and I could hear plenty of cursing going on.

We had to go hand over hand. Then someone said this pile was here from previous floods, and now its flooding again! Jeez it was really panicky! After a while we reached the rocks. Must 'a bin the wall of the ravine but the climbing became even harder 'cause we had to cling to little cracks and ledges. It was so bloody dark we couldn't figure what was up or down. And the river just bloody well roared! Never heard anything like it!

All the time the boss was reassuring us and getting us to yell out so that he knew we were still there. Anyway, in the end the river seemed to stop chasing us, so we stopped climbing and just sat there perched on a few little ledges. It was a really cold night and we were soaked to the skin and shaking. The rain wouldn't stop. It was hard to figure out where everyone was. Don't know how the boss did it!

Anyway, in the end it got lighter and we were able to see where we were. Three of the rafts had gone and the boss was really cranky because the 77 set was gone. He was going to get it for that! The tents that some of the blokes had put up on the logs when we first arrived were gone, with all the gear!

All day it rained! Then the boss got us to keep climbing up to a small ledge that he had spotted, 'bout another 15 metres up. We were able to stretch out a bit there. Didn't know how he was going to get us out of here. Be impossible to walk out I reckon. Just climbing up to the top was probably impossible anyway! Then, just as it started to get dark again we heard the chopper. Shit what a great sound that was! It was flying above us somewhere, but we couldn't see it cause the clouds were so low and the whole ravine was misty. Luckily the boss had managed to save a few flares. He fired one. It disappeared! Then the chopper must've come back 'cause we heard it circle above us. The boss fired another flare and the pilot must have seen it because he just kept circling.

Anyway, we all knew it'd be impossible for him to land on the side of a ravine, but he knew where we were now, I reckoned. We had another crappy night. Some of the blokes had managed to rescue their packs and they put on some brews and got some food going and we all had some. Some of the injured blokes were starting to get real sore. And it was so bloody cold! Next morning it was still cloudy, but we heard the chopper looking for us again. We figured he couldn't land again 'cause he went off again. Everyone was really pissed about that but later in the afternoon the cloud lifted and the pilot came back, would you believe?

The boss'd worked out how to get us to a ledge that was higher up the ravine. Only one at a time! He figured the pilot might be able to land there. He arranged for the two most injured blokes to be on the ledge first, ready for the chopper. The boss threw some smoke so the pilot could find the ledge and we watched as he put one skid on the ledge. Couldn't believe it—the other skid was hanging in mid-air! It looked bloody risky. There wasn't enough room for the blades to spin if he had put both skids down. Jeez he was a skilled pilot.

The boss leaned into the chopper to talk to the pilot and then the first two blokes loaded, and the chopper took off. The boss said the pilot was going to take them to Strathgordon. Then he

This is where Captain Ian Johnson's Group camped in the Great Ravine, perched on logs with little space to retreat up the sides of the ravine when the water rose later in the night.
Photo Gunner Tony White

came back about an hour later, and it was Peter Nolan's turn. It looked like he had broken ribs. The boss said that Peter would be taken to the end of the river RV, where we were all supposed to be going, because the other groups were already there.

The rest of us would have to wait here another night. Pilot didn't have enough fuel. He'd be back at first light. We were pretty pissed off with another night here, but what can you do? He did come back, and he took two of us to Strathgordon. We were going back with those guys by road to Hobart. The boss stayed there a bit longer and the chopper was going to come back and pick him up and take him straight back to Hobart.'

A gunner's account, taken from first-hand accounts from Group 3, 12 Dec 81

There is no doubt that it was a lucky escape. But luck is often the result of self-help. The Army and its insistence on methodical planning and training got them through, saved their lives, and saved my skin, and my career! I would have some serious explaining to do to in Canberra about the loss of a radio, but that was a worry for later. The loss of a life would have been an entirely different matter. I shuddered to think about it. Harry and the pilots had had their wits about them and as the rain got heavier and heavier, they figured that there was a good chance that at least one of the groups might be in trouble. They made a plan. As soon as the weather allowed, they would fly the length of the Great Ravine, just to check, in case they might be needed. Brilliant!

On the afternoon of the second day that Johnno's group was on the ravine wall, the pilot thought the weather had cleared enough and set off to see what may be happening. He couldn't see the river because it was shrouded in mist, so he flew well above it, following the river's course using his instruments. It was Johnno's quick thinking that saved the day. The helicopter just happened to be in the right spot, and the pilot just happened to be looking in the right direction, when the flare broke the cloud, bursting in front of his eyes! He threw the aircraft into a steep turn and within a few moments a second flare broke through to confirm what he saw the first time. There was no answer to his radio calls, so he did a few circles, just to let those on the ground know that he had spotted the flares, marked the location and flew off, ready to come back the next day.

I didn't know this until much later, as bits of information trickled in, and after I could have a good long chat with Ian. For now, however, I knew they were all safe... *we* were all safe, and the trip was in wind-down.

My group, together with Mal Booth's, would meet the Denison Star the next day. Here is how Don Johnson saw it:

> The groups that had finished the river met the tourist launch Denison Star in the lower Gordon. Tourists on the daily run were treated to a display not unlike the boarding of a passenger craft by river pirates, as 14 yellow rafts swarmed out from the lee of Butler Island and boarded them from all sides. I struck up a conversation with an elderly deck hand... who had signed up for war in 1939... and an elderly passenger who had also served in the last war. What were they like they asked me,

these gunners standing in happy bibulous groups, each with its attendant revenue of fascinated tourists? I assured them from what I had seen, the metal was still as good as it had ever been. Adventure is what they had come for, and training, and they had received both in full measure.

The passengers continued to appropriate groups of soldiers. I later struck up a conversation with Terry the old barman. He worked on Huon pine cutting in the 1930s. Rowed an 18' punt from Strahan up the Gordon with one other and chaff aboard for the horses.

The soldiers are drinking cider. Far to the east is Frenchmans, shrouded in cloud... the atmosphere on the boat is memorable. The first time we have seen the reaction of the general public to our trip. A sudden translation from discomfort to warmth and luxury. Mild contempt for the pallid civilians.

Journal Entry, Don Johnson, 10 Dec 81

By the time we reached the wharf at Strahan, we were well and truly under the influence of the cider and beer we had been scoffing, mostly shouted by the tourists. We felt like celebrities! We unloaded, slowly and deliberately. The dock was empty, especially after the tourists were bundled into their buses and found their cars to drive to their comfortable accommodation. But there was tension in the air. The animosity was palpable. Contempt poured out of the pub across the road, directed straight at us.

In an instant, we went from celebrities to outcasts. Pretty Strahan: picturesque; rich in European heritage; gloriously colourful; full of Tassie rednecks, Hydro workers and wood cutters, men for whom we were apparently the enemy. They took us for a bunch of greenies! We must have looked an easy target because a couple of them stole an ammunition container that was sitting on the wharf, that had just been unloaded. It was an understandably tempting target, but the thieves had no idea of what they had stolen. It was actually one of the group's toilet boxes, filled with plastic bags full of the combined faecal matter of seven blokes on hard rations over 12 days! A collection not to be envied and a worthy prize indeed for such a daring raid. That night we had a few more beers to top up the start made on the ferry, and I went to our shared room and fell into a deep, happy, dreamless sleep.

The next morning, we headed off and away from the west coast. Sitting on the bus the next day, on the way back to Hobart, deep in thought about the wonders of the trip, looking forward to seeing my family, and feeling that tinge of sadness that seems to be the trademark of mine when leaving the wilderness behind, I felt a tap on my shoulder. Turning, I saw Tracker as he pushed a handful of money into my hand.

'That's $250 for battery funds boss', said Tracker. Then silence.

'Okay, so what gives, spill the beans mate', I said.

Then the story came out. In the night, long after I was fast asleep, it seems some of the local redneck boys thought they would teach these greenies a lesson, and after consuming a skinful, some of them took it upon themselves to use knives to slash a few of the tents. Then, like the heroes they were, they sped off into the night. What they had failed to realise was that these greenies were in fact Army boys, and not just any old Army boys, but gunners, used to dragging heavy howitzers through the mud and lifting ammunition all day. Apparently, the battery boys reacted quickly.

'Harry got Biggsy and Barry and a few us together and we followed the noise then we felt the bonnets of the cars up on the hill and we figured that we had them! The lights were still on inside this house, so we knocked on the door and told them they owed us for the damage! You should have seen the looks on their faces! Harry was very polite, boss, so don't worry!' Tracker said.

You can imagine the surprise for a group of these local lads, when they were confronted with some of the biggest men in the battery. These gunners had spent many years of Army training dragging big howitzers around, and were the types that you don't cross. Tracker, who was a part of this group, and backed up by Ian Biggs, another participating miscreant, told me that the gunners just stepped into the house, one after the other, until they filled the lounge. Harry, my jack-of-all-trades Sergeant, was apparently the ringleader. I had guessed as much—a man of many talents, is my Harry. In a polite voice and with a little smile, as is his trademark, Harry calmly suggested that the locals may like to consider some form of monetary compensation. Apparently, the blokes in the house capitulated immediately. What else would you do if you are faced with a lounge room filled shoulder-to-shoulder with Army gunners? The local boys, after being gently reminded of their misunderstanding as to the identity of their visitors, reached for their wallets and the money was handed over. Not a hand was laid on anyone. And all this whilst I was blissfully fast asleep!

I must admit, though, I was uneasy. What if these locals were to go to the police? I could just see the headlines... 'Army extorts defenceless locals'!

'Onya Tracker... well done!' I said.

The end of the trip, only one member of Group Two made it this far, the rest of Group Two having been ferried back to Hobart from the Great Ravine. Rear from the left Major Ben Lans, Corporal Mark Lowrie, Bombardier Alan Ward, Lieutenant Mal Booth, Sergeant Norm Wheeler, Mr Herbert Van Daalen (photographer), Sergeant Vic Shields. Front from left Gunner Ian Boyd, Bombardier John Webb, Bombardier Ian Biggs, Sergeant Peter Nolen (only member from Group Two), Gunner Tony White, Gunner Graeme Jorgensen.
Photo Don Johnson

Oh well, I thought, we should just enjoy the ride home.

I kept that money and added it to a donation that the gunners wanted to make to the Tasmanian Wilderness Society to help in their fight to save the river. It was accounted for with an entry of... 'Donation to Save the Franklin on behalf of concerned locals'. How apt that these red-necks contributed to the saving of this great river. They never knew.

22 THE AFTERMATH

Back in Hobart, the weight of responsibility was off my shoulders and it was time for a few drinks with the boys before we caught the transport home. I'd built in a bit of spare time, in case the river trip took longer than anticipated and I couldn't change the date of the RAAF Hercules that was to take us back to Townsville, so there was nothing to do but pack the gear and have a good time. We had a responsibility to our sponsors, so I spent a bit of time with Herb choosing photographs for printing and sending to Minolta, Primus, Kodak and other sponsors.

Don phoned his newspaper contacts and started to write the copy they wanted. They all agreed to give him a 'by line'. The articles were well written and described the wilderness, the solitude and beauty of the Franklin. He was accepted by *The Australian* Newspaper for a double page spread in the Weekend Australian, which was a feather in his cap and great publicity for us, the Army and the Franklin River.

As the time grew near for departure back home, I became saddened by the thought that the adventure was finally over. Not only was this trip over, but I was about to leave my best, most enjoyable posting in the Army to date. My tenure as Battery Commander was complete. In January, a new Battery Commander would take the reins. This is normal in the posting cycle of the Army, but I would have done anything to go back to Townsville with the boys, just to round off the adventure that we had just completed, gather all the stories and make sure everyone was okay. After everything that we'd shared, not just on this river adventure, but during the years of training: the hard work, the fun, the failures and the achievements, it was ending, well for me, anyway. They were going back, and I was moving on.

On the last night in Hobart we had a huge celebratory dinner at a restaurant, and then we caught the Herc home. I left the flight in Sydney where I had a wonderful reunion with Lo and the kids. Lo was so excited by the stories of the river that right there and then, we decided to raft the river together next year. I was convinced that this remote, wild river flowing untamed through the raw nature of south west Tasmania was unique in the world. I wanted to be there when Lo saw it. She loves the wilderness as much as I do, and this was her right. In the previous three years, whilst I was enjoying the best posting of my career, she was keeping the family going. This was going to be something special for her. There would be no helicopters next time, and no support crews, but I now had a wealth of experience with which to tackle a second trip. Besides, the thought of being back on the river in a year was a good way for me to soften the pain of leaving it now[3].

There was still a lot of work to be done. I decided that we needed more publicity, so I joined Don in writing articles for newspapers, accompanied by plenty of colourful pictures. Getting published turned out to be the easy part. Just at that time, apparently the whole of Australia was becoming aware of the Franklin and the plight of the river, with the Tasmanian Government hell-bent on damming it, and the environment movement hell-bent on saving it. Newspapers are all about news and an Army group having an adventure on this remote and challenging river, a river which at that time had rarely been visited,

---

3   As well as the trip that Lo and I completed the following year after the adventure training event, we rafted the river again in 2001 with our three children Matt, Rae and Rob and their partners. This time not in rubber duckies, but in large four- and six-person rafts. Matt, Rae and Rob heard so many stories about the river when they were young, they made us promise to do it with them when they were old enough. Matt was the one who announced, one day in 2000, that it was time! It was a huge family achievement and a great trip.

was news. The difficulty however, was to ensure that the sponsors knew when each of the articles were to appear, so that they would have an opportunity to place an advertisement alongside the article, boasting of the equipment that they had contributed.

In addition to the *Weekend Australian*, which called their feature 'The River of No Return', we were published several times in the *Army Newspaper*, the *Australasian Post*; *Wild Magazine* (although not as a major feature); the *Townsville Bulletin* as a major multi-paged Saturday feature entitled 'Army's Most Ambitious Training Exercise'; Brisbane's *Sunday Mail*, with a three-page coloured spread; various product magazines and the *Hobart Mercury*. The *Mercury* called us a 'crack' Army unit and gave us a great wrap. We were actually doing something on the Franklin, just as the rest of Tasmania was preparing to vote in a referendum on the future of it. The publicity we received caused other commentators to mention us, in columns like 'Clancy of the Overflow', which was published in the *National Times* and where the author commented that the Franklin might be a suitable military training area!

Ironically, I found myself to be a reluctant hero one day, whilst sitting in a barber's chair in Brisbane. The barber had just read the *Sunday Mail* feature and, being quite excited by it, proceeded to tell me all about it. After a while he figured that this particular head of hair that he was talking to, knew a little more about the topic than the average head of hair he cut. Suddenly, he recognised me from the pictures in the paper.

'You're actually that bloke, aren't you? You're that Major in the Army that led that expedition, aren't you? Yes, you are, I recognize you!'

And that was it. He called all his mates from shops up and down the street, and they all wanted to shake my hand. That was my moment of fame, in a barber's chair! Grinning faces watched my locks being trimmed by a barber who chatted incessantly, resulting in a very short haircut!

I was proud of what we had achieved, and that we may have helped, just a little, in bringing the plight of the Franklin River into the spotlight. Of course, I was also happy that this exercise had helped to highlight the benefits of adventure training in the Army. These words, as part of the *Weekend Australian* feature, appeared:

> The unit which made the trip is something of a pioneer in the field of adventure training: the 4th Field Regiment's 108 Field Battery, is the Army's Operational Deployment Battery, slated to be first into action in the event of Australian military involvement. They are understandably trained to a high pitch of combat efficiency, as they demonstrated in the recent Exercise Kangaroo 81.
>
> But training alone does not prepare a unit: a more precise test is needed to try the courage and the initiative of the individuals who comprise the unit. The urge to face danger is part of a soldier's make up, so the Australian Army has instituted adventure training, pitting soldiers against natural dangers.

Weekend Australian 6-7 March 1981

But now I was no longer in 108. I was now posted to Brisbane to be part of the Headquarters Divisional Artillery. In my job as a part of that headquarters I would travel Australia conducting technical tests and assessments to determine the standards of training and readiness of the Army's Artillery Regiments. In this role I would soon be going back to 108 Battery, but this time to assess their competence, not as their commander. However, it was during this posting that a few 'post-activity' problems surfaced.

I had taken a huge career gamble in soliciting for commercial support, or 'sponsorship' in the vernacular of expeditions. The permission that I had sought, and received, was limited and I knew that I had exceeded the spirit of the approval.

Clearly, now that the exercise was over, I could no longer push the issue aside. There was a need to do a bit of 'behind the scenes' ground work within Army circles and warn them of the avalanche of publicity that was about to hit the mainstream press, just to cover my back. In Army Headquarters, the Director General of Army Training (DGAT) had directed me *not* to infer Army endorsement of *any* product. Also, in their ignorance, or perhaps naivety, the Army expected the *Army Newspaper* to be the only publication that would be sufficiently interested in carrying any articles on our trip. This contradicted the expectations of the many sponsors that I had. They needed publicity, tons of it, and they needed my words of endorsement.

I played this very carefully and was always guarded in my words, or so I thought, and accompanied any potential endorsement with a disclaimer that this was not the Army's official position.

I was nervous. I had informed all the sponsors of our various projected publication dates, as per agreement. Sponsors like putting their advertisements next to any activity they sponsored. That's why they do it! Let's face it, the Franklin River was the flavour of the time! But I was nervous because I knew the sheer volume of publicity and frequent correspondence with sponsors made it almost certain that something would go wrong. There were so many articles appearing all over the capital cities of the east coast, unmissable because they were often multi-paged, accompanied by headline grabbing front page quotes. Someone in Army Headquarters would surely see or be alerted to what I done, and realise that I had stretched my authority.

Sure enough, it happened. *The Weekend Australian* feature was published: a massive spread over two huge pages, with sponsors' advertisements all around the edges of the accompanying photographs. This was the big one. Naturally, since *The Australian* is a national newspaper, the feature was seen Australia-wide. Before I got the call from Canberra, I decided to 'self-report'. Better to 'fess up than try to act innocent. I had a mate in DGAT, a Lieutenant Colonel whom I trusted. I warned him of the publicity that was out there and just how much sponsorship I had accepted, and that the proverbial 'shit was going to hit the fan' shortly. He was taken aback.

'Mate, this is going to be ugly if one of your sponsors does the wrong thing and mentions the Army as a promoter of their product. But I will try to look after you', he said.

It happened! It got ugly! I'd accepted an offer from Hallmark International, the manufacturer of rainproof equipment called 'Z-Kote', to try some of their products. They made weatherproof coats and trousers in direct opposition to Gore-tex, and they had offered us three sets. I should've knocked them back at the start, but my inexperience with industry showed, and I was tempted by their smooth talk, their assurances that they just wanted some idea of how their kit stood up to the rigours of the south-west of Tassie— they only wanted some 'personal' opinions. I and two others had used the gear on the river. It was good. In fact, it was very good. I said so in my report to their Hallmark General Manager. But then the General Manager came back and asked me how it compared with Gore-tex, and that is where I made my mistake. I said it compared favourably and I mentioned both products in the same sentence. This was not in writing, but a phone call was enough for him!

Hallmark published a version of my words in their internal magazine and added some imaginative descriptions of their own: 'Test Report by a Research Group from the Australian Army', they called it. This sank me. Their internal magazine, with my words of favourable comparison, were passed, or should I say 'leaked', through industry sources, to the makers of Gore-tex, who promptly submitted a formal complaint to the Army. Why was this so-called 'official test' sanctioned and why had they not had the opportunity to offer their products in competition? I had been set up by Hallmark.

The complaint travelled all the way to the Chief of Personnel—Army, Major General B. H. Hockney, but I was only partly aware of what was happening. At Army Headquarters in Canberra, my friend in DGAT was doing what he promised. I had just about written the issue out of my mind when unexpectedly, nearly a year later, I received a letter from General Hockney. I opened it and with a shock instantly understood how close I had come. The fact that I was not called to explain myself to Canberra meant that my mate in DGAT kept me from the flack as much as possible, and had been explaining things on my behalf, taking the venom out of the situation by pointing out that I had tried to do the right thing, but that I had been set up by a company that simply wanted publicity.

I was severely wrapped over the knuckles. Here is an extract of the letter sent to me by General Hockney:

> The approval sought (by you) and given (by the Army) was quite explicit. You exceeded your authority and as a result placed the Army in an embarrassing position. I recognize that some steps were taken to comply with the spirit of the (Army's) direction, if not the letter, however I expect an officer of your rank and service to anticipate the consequences of your actions and to appreciate that an authority[4] to 'state quite freely how and by whom the report was originated', could be manipulated and construed so as to imply some form of endorsement by the Army.
>
> In all circumstances, I have decided that it is sufficient to point out your shortcomings in the handling of this matter rather than formally censuring you. Accordingly, I have decided it would not be appropriate to place a copy of this letter on your dossier in the Office of the Military Secretary.

Letter from General Hockney, 15 December 1983

Phew! That was a relief. Getting kicked in the bum is all part of pushing the limits, but had the General informed the Military Secretary, the consequences to my career would have been serious.

I told myself that I had tested the limits, and won, well sort of, anyway, depending on who's point of view was taken. Major (later Lieutenant Colonel) Brian Agnew, a friend and colleague, who was a prominent member of the AAA and who was in the process of helping to plan the AAA's 1988 Mt Everest Expedition, personally thanked me for opening the door on

---

4   General Hockney assumed that I had given Hallmark permission to use my name which was incorrect. I believe this information may have been volunteered to the General's staff by the managers of Hallmark, in an attempt to deflect the situation and prevent the Army from taking legal action against them. As the whole conversation had been by telephone, there was no record of what I had actually said or given permission to use. I was therefore in no position to argue as it would have been their word against mine.

commercial sponsorship. Not long after my Franklin trip the official policy was changed. From then on, Army expeditions could accept sponsorship under a more relaxed set of rules. Phew again!

One more episode which occurred as a result of the trip is worthy of mention. In the last few days before departure from Townsville, I realised that I was short of cash to finalise some of the payments. I had arranged for the soldiers to make contributions to the overall cost of the activity which, in spite of the sponsorship, was still substantial. Not all the money was collected in time, however, so I asked the CO if I could borrow $2000 from Regimental Funds, to be paid back next year. He agreed. As I was a signatory on the fund, I wrote out a cheque to myself, cashed it, and thought no more about it. I would square the books as part of the exercise wind-up. By January I was no longer a member of the regiment, and as it happened, I was sent on a promotion course where I found myself studying tactics and strategy instead of worrying about finalising the exercise books. Exceptionally bad timing! Within a day or two of starting the course I received a call from a Lieutenant Colonel Jim Ryan. He was the new CO of 4 Field Regiment. His words were icy:

'Ben, I see that you have taken $2000 from Regimental Funds without any authority or mention in the minutes. You have 48 hours to repay the money or I will personally see to it that you are cashiered out of the Army for fraud and theft!'

Whoa, my throat closed up as I tried to speak. What the hell? I stumbled with an explanation that I had indeed sought authority and should he just call the previous CO, Lieutenant Colonel McGuinness, he would back me up. But I was in a weak position. Not to record the approval in a set of minutes, countersigned by McGuiness, had been my rather stupid and obvious mistake, so keen was I at the time, just to get going on the exercise. It was another lesson for me to tuck away.

Transferring that much money in the days before electronic banking was not easy, particularly whilst being away on a course. I rang Lo and she was tasked with finding the money and sending it by express transfer. This was a matter that I should have taken care of, but my trust in people is such that I always expect others to treat me in the way I would treat them. Surely this officer didn't think that I would try to defraud the regiment of a paltry sum of $2000, and put my entire career at risk, by simply signing a cheque to myself? Later that year when I came to Townsville and put 4 Field Regiment through a demanding test routine, as part of the Divisional Artillery test team, I did my work, gave an honest assessment, looked Colonel Ryan in the eye and never mentioned a word of what happened before, and neither did he. We moved on.

The Franklin and its aftermath was finally over. Would I do it all again? You betcha!

23 A STAFF POSTING AND STAFF COLLEGE

The posting to Divisional Artillery Headquarters in Brisbane in 1982 was my first posting to my home town and it was a timely one, as my parents were getting older. It was nice to be near them, and for the children to get to know their paternal grandparents a little better. When we moved to Brisbane, Robbie was just a little fella, Rae was about to start school and Matt was just transitioning to Year One. I was posted as a staff officer on the Divisional Artillery Headquarters, situated at the huge Army base at Enoggera.

The job was a 'staff job'. I had to complete a staff job in order to get that important tick in the box as an all-round, well-trained officer. We managed to find a house to rent in Nudgee, not the most auspicious Brisbane suburb, but it was in a nice street and it was relatively close to Enoggera Barracks. It was also close to Aspley, where my parents lived. Besides, it was one of the few rental places that would allow dogs, so we could take Jo, our German Shepherd, with us.

Life at the headquarters was dictated by routine, rather than innovation. Everything was driven by the training calendar of the division, which meant one large-scale exercise every two years plus a few smaller 'command post' exercises. My friend from the old days at Portsea, Hugh Polson, who was also posted there, decided one day that the place needed livening up. His choice of methods to achieve this 'lively' state was one that only Hugh could devise. It had his trade mark stamped all over it. Every week, the Divisional Chief of Staff (CoS), who was a full Colonel, would hold a meeting. At this meeting, all the headquarter branches would send their representatives. The room would be full of Lieutenant Colonels and Majors from branches like operations, logistics, personnel, engineers, artillery and so on, who would wait diligently for the good Colonel to arrive. He was a particularly dour man. Large framed, quietly spoken and with no discernible sense of humour, he was a man who took his job far too seriously. Hugh, who saw life as a series of fun things to do, took this as a challenge!

In his inimitable way, Hugh decided that the CoS, whose birth date he had just discovered, and which coincided with the next weekly meeting, was to be the means by which he would get a bit of cheer into the place. Not being one to do things by half, he arranged for a 'gorilla gram' to be delivered at the weekly meeting, to none other than the CoS himself. Picture a room full of serious looking officers sitting around a large boardroom table, waiting for the senior man to speak, when the door bursts open and a man dressed in a gorilla suit enters, jumps up on the table and, to the amazement of everyone in the room, starts dancing and singing happy birthday to the big, important looking man at the end of the table! After a moment of stunned incredulity, the room explodes into laughter, the meeting breaks up in disarray, the gorilla escapes, 'whooping' down the corridor and out into the street, and all the headquarter staff who have not made it into the room but, having been warned by Hugh, are waiting outside, start cheering and singing 'happy birthday dear Colonel!' There is raucous approval! Hugh is a star! But wait... is he? The Colonel remains unmoved, grey faced, and still sitting in his chair waiting for the mirth to die down, so he can get on with his meeting.

Not even a gorilla-gram could move this man. It moved Hugh though. His ears were ringing for days. The man was not amused, but Hugh's reputation as a fun-loving officer was immeasurably enhanced.

The Divisional Artillery Headquarters had a small staff of command post operators, signallers and drivers, about 20 in all, who were there to help set up and manage the tests that we conducted. As a posting, being a soldier or NCO in the 'Div Arty Troop', as it was called, was fairly unremarkable. They had few responsibilities outside the actual work they did in the field assessing the artillery units, and their day-to-day barrack life was quite dull. So they decided that, if this adventure training thing was good for soldiers in the field units, it was okay for them too! I was shocked when I began the posting to find that the whole troop was waiting for me to arrive so that I could take them all down a river somewhere. My reputation was preceding me. My boss, Lieutenant Colonel Denis Casey, was happy to let me do as I wanted with the troop, so I decided to plan some trips for these boys and get them... 'off their arses' as the Army saying goes! Once again, I ordered little rubber duckies, this time for the artillery troop.

I selected the Snowy River as a suitable river for adventuring. Bad choice! It was a long drive from Brisbane to the Snowies and when we arrived we were disappointed to find that there was little water flow. It became obvious that I had not done enough homework on this river. In spite of the Hydro Authority's assurances that all would be well, the people I spoke with were not river rafters and had no idea how much water was needed to make the river challenging. We did have some fun but struck only one challenging rapid on the entire five-day paddle. I felt so bad for the soldiers that I immediately promised to take them on a more exciting trip the next year.

In my attempt to remedy the situation and give these blokes some excitement and variation to their routine environment, I went from the sublime to the ridiculous. The Nymboida River in northern NSW is one of the great white-water rivers on the east coast of New South Wales—not a river for beginners. It's not suitable for little rafts because, despite the numerous rapids, the stretches of flat water are frequent and long, and the trip would take many days of paddling in flat water: a prospect not to be relished. So, we needed white-water kayaks. Two problems. We didn't have any kayaks and nobody, including me, had any idea how to paddle them even if we did. The level of expertise needed to negotiate rapids in a kayak instead of a rubber raft takes more than just a few weeks of practice. Nevertheless, the troop was not deterred. I had promised them adventure and after the flop of the first attempt, adventure they were going to have!

'Your boys made their own canoes in Townsville, didn't they, boss? Why can't we make our own kayaks here?' asked Ivors Valuks, the Sergeant in charge of the troop.

'Errr, well,' I stumbled in my attempt to say something.

This was not what I was thinking, but I could see the enthusiasm in their eyes. Somehow, I was implicit in adventure training and these boys wanted a slice of it.

'Okay,' I heard myself say. 'I'll work something out. You blokes get the sheds at the back of the headquarters ready for some fibreglass building work.'

In no time, we set up another production zone, this time for double white-water canadians. In my misguided thinking I decided that it might be easier for the boys to paddle double kayaks rather than single ones. So we made banana-shaped, banana coloured, hybrid double kayaks with a hatch each for two people, and a bend that made it pivot and wobble in the middle. They were not the most graceful looking and, in the water, they were a complete dud! Easy to turn all right! So easy that the boys had trouble identifying the 'front' and 'aft' of these weird boats. When we were on the river, it was not unusual for any two paddlers to be seen comfortably and confidently heading downstream, backwards! But it mattered little. No one seemed to worry about the fact that we were about to tackle some serious rapids with crazy yellow boats that we couldn't handle! We headed for Grafton to camp at Goolang Creek, a tributary of the Nymboida, where we established a base to give ourselves some time to figure out what made these yellow bananas work on the water. Their moment of adventure had come! The plan was to set off down the Nymboida to the junction of the Clarence, past the Clarence River Falls (which we learned had killed a father and his son in the last few months) and on to a rendezvous point downstream, near Grafton.

Carting the canoes over sharp rocks beside the Nymboida River.
Photo Author

The exercise was a technical disaster, but a roaring success. Lots of good old Army-style bonding around the camp fire. The paddling was hilarious. Most of the men tipped out on most of the rapids. I have rarely seen blokes look as uncomfortable on the water, with their knees sticking up out of the cockpit, paddles flailing! Fortunately, as we progressed down the river, they became marginally more used to the craft. I must admit that the design of the canoes was not great. There was no clear centre line, just a flat bottom, rounded out at each end, resulting in a bow shape that would swing wildly every time someone put a paddle into the water.

The Clarence Falls, steep and treacherous, were well known for their deceptive approach. We had been warned to be on the lookout for what was described as a 'false horizon'. Experienced Clarence River paddlers told us to watch for the appearance of a line across the river, an illusion really, above which things would suddenly appear to seem more distant than they should. This was difficult to imagine, but when it happened we saw immediately what they meant. A line! There it was, a sharp demarcation of 'near' and 'far'.

Two of the group attempting the Clarence River in self-made yellow banana shaped whitewater canadian canoes. The inexperience of the paddlers is evident in their paddling positions and style and they were tipped out shortly afterwards.

Photo Author

At that point the water just dropped away with no sound because all the noise was projected forward. No warning ripples, as there were no big boulders or trapped tree trunks to back the water up, forcing any sort of surge! The current was the only warning, increasing imperceptibly until suddenly we were aware that in front of us the river surface dropped away and we were staring at the new river surface, about forty metres down.

We portaged the falls via a fast running narrow side chute, where a portion of the water barrelled along, dropping quickly over rocks in a channel leading down to the new level below. It was perfect for roping the kayaks down. Unfortunately, I was called away in the midst of guiding my kayak with my paddling partner, a Captain Dave Morgan, who was from a different part of the headquarters.

'Okay, mate, stay at the back of the kayak using only the rear rope to guide her through. If the nose gets caught behind an obstacle just pull the rear end back gently and let it find its own way. It'll generally follow the greatest volume of water. Don't try to push from the nose because the kayak will swing immediately. It'll tip! It'll fill and snap in two. Just like that!' I said, snapping my fingers.

'No worries,' was the answer.

When I came back after just a few minutes, sure enough, my kayak was in two pieces!

'What the...?' I said, frustrated.

'Sorry mate...' he began.

I was deflated.

We cobbled the two halves back together with 100 mile an hour tape and struggled on. I gave him such an earful that he has never forgotten it. We were both out of the Army by the time I saw him again, some 25 years later, and the first thing he did was apologise again! Was I really that hard on him?

A few days later we headed back up the highway to Brisbane with a load of busted and broken kayaks. I am not sure if anyone ever used their banana boat again, but everyone got their adventure and they were happy.

This is the team that tackled the Nymboida. A mixed bag of soldiers with little or no experience but determined to have a great time.
Photo Australian Army

The other major event in 1982 was the second trip down the Franklin, a year after the 108 Battery trip. Lo was chaffing at the bit to go. In addition to both of us, Mal Booth, Ian Biggs, who had been with me the year before, Ian's wife Dianne, and my nephew Glenn, made up the party. At the last minute, we picked up a stray paddler just before setting off from the Collingwood Bridge. David Smurthwaite, whose father was a doctor in Melbourne, had decided to do the river all on his own. David was unlike any doctor's son I could imagine. Indeed, he was unlike any river rafter! He wore no shoes, had only one set of clothes, carried a huge string bag of vegetables as his only food for 12 days, possessed no map or had any knowledge of the river, had never rafted before, and rarely spoke. But he was very likable, so we asked him to join our party, which he did gratefully, saying that he had hoped to bump into some other group doing the river. He turned out to be quite an asset. Strong and willing to do anything, he quickly worked his way into the group, particularly after a few days when we, having no vegetables at all, started to hang around behind him as he carried his huge string bag about the place, in case a carrot or two dropped out. He did sometimes share his vegetable soup with us, which we devoured.

Once again, the Franklin was glorious in all its isolated splendour. We managed to find the cave that had eluded us the year before: Kutikina Cave, where man had lived during the last ice age. Unfortunately, the area was alive with tiger snakes. We saw at least ten of them in the strip of land between the edge of the river and the cave. Dianne, Ian's wife, became so scared that she actually appeared to be going into some kind of melt down, so Lo volunteered to take her away downstream to a safer spot. This selfless act meant she missed seeing the cave.

That river did something to Lo. It came to life in her heart and as we entered the Gordon at the end of journey, she had tears streaming down her face. Some moments in your life are so special that they stay in your memory and remain as fresh as if they are happening now.

In December 1983, at the end of the second year of my Brisbane posting, we packed up once again and set off to Queenscliff, in Victoria. Queenscliff was the location of the Australian Army Command and Staff College.

Something that I always enjoyed in my Army career was coming home to Lo and announcing the next posting. She'd become accustomed to moving at least every two years, and sometimes more often, usually to an entirely new place in a different state, or even country, and she enjoyed the challenge! She loved the idea of a new place to go, in spite of the general uprooting of everything. Lo has adventure in her heart, just like I have, and the two of us would look forward to the 'new'... the unexpected. However, the news of another *overseas* posting added another level of excitement!

I had been agitating for an overseas posting, and, just to stir the pot, had been telling all my fellow officers that I was being posted to Malaysia. This was not true of course, but after Singapore, Lo and I fancied another Asian posting. After a few months of this, however, I received a call from the Artillery Directorate in Canberra. It was my good friend Major John Derbyshire, who was responsible for officer postings.

'Ben, I hear you think you're going to Malaysia. I don't know where you got that from mate, but you're mistaken, you're going to England!' he said, clearly happy to deliver this good news.

Well, there you go I thought. Perhaps my agitation did the trick. Now, more than a year out, I already knew my next posting, even before attending Staff College, a place that was famous for exerting lots of pressure on students, on the pretence that the best-placed students would receive the best postings.

We made a bit of a holiday of it on the way from Brisbane to Queenscliff, camping with Lo's mum and dad, Gwenn and Cliff, plus her sisters Lyn and Leonie, at Tom Groggin, on the upper reaches of the Murray River. We wanted to use our Franklin rafts to do the rapids in what is known as the 'Murray Gates'. That was an eventful trip. Not exactly a relaxing family holiday. It rained and rained until it rained us right out of the place into a rental house at nearby Khancoban, where we played cards and board games, safe from the rain.

Before the rains came however, I managed to take Cliff, Lyn and Leonie and their two kids, aged around ten, plus Matt aged seven down the gorge with its grade six rapids. Apart from kids getting bashed over the head by our paddles a few times, and Cliff getting dragged under a tree by the current, potentially an extremely dangerous situation, everything went really well. Mostly, anyway! Somewhere upstream there had been a sewage contamination of the Murray, and soon we were all as sick as dogs. During the incessant rain we were tent-bound, weak from the incessant 'runs'. I had to dig a trench close to the tents so that the stricken didn't have to run far in the rain to aim and squirt.

However, there's no keeping Lo down, and whilst the family was ensconced in the house at Khancoban, we managed a successful descent of the gorge in flood, without too many mishaps. It was a fast trip down a swirling river straight over the grade six rapids which we dealt with easily until Lo became stranded on a large rock in the middle of a rapid. She had tipped out on a tricky boil, quite easy to do in those silly little rafts, and, having found safe ground, refused to go any further. I could understand that! I scrambled back upstream with a li-lo and managed to deliver it to her and she took off down the current, hanging on for dear life! But predictably, I was now caught on the same rock. I decided that swimming the rapid would be oaky when actually it wasn't! The current immediately grabbed me and pushed me into a sheer wall, where the river, in the process of doing a ninety-degree left-hand turn, pulled me straight down and sucked my wetsuit boots clean off my feet before letting me go. I learnt yet again never to underestimate fast-flowing water.

It was an eventful trip and when the rains finally stopped, we headed down to Queenscliff whilst the rest headed back north to Sydney. Billy Foxall, an Army mate from the early 4 Field Regiment days, was my sponsor and had arranged our rental for us, and what a piece of luck that was! We moved into a beautiful little house, over a hundred years old, comfortably renovated, about 150 metres from the front door of the college!

I had mixed feelings about attending Staff College. My 'head' still wanted to be at the 'soldier' end of the Army, helping to manage soldier careers and watching out for their mental and physical well-being. I wanted to remain in postings close to men, where face-to-face leadership counted. Here at Staff College, the emphasis was on higher-level planning, preparation and strategic development. Later in my career, I came to understand the lessons taught at Staff College. But for now, I wasn't ready.

A significant portion of the work was our major Staff College paper. This was to be a 20,000-word treatise on a subject of our choice, as long as it fell within the broad categories of training, military history or military strategy. Being a keen advocate of adventure training and wanting to demonstrate to the Army the relevance and benefits of that kind of training, including the need to embrace adventure training as an accepted training regime to be adopted by Army units, I chose as my topic: 'The Benefits of Adventure Training'. All topics had to be approved by the Chief Instructor, but he was not convinced.

'A bit frivolous, Lans. I will approve the subject but why don't you choose a REAL topic!' He said with a smirk.

Obviously, he was an adventure training 'nay-sayer'. I should have seen the writing on the wall. If the Chief Instructor didn't like what I was writing about, perhaps I should have taken the hint and chosen a different topic. But, he was the kind of senior officer I needed to convince, so I politely refused his suggestion to change topics and wrote my paper, promoting the idea of establishing an Army Adventurous Training School.

Quite predictably, the comments from the college were non-committal. This was especially galling, because it appeared that others *did* value my paper. In those days, before the internet, the University of New England was offering a course in leadership training; they heard about my paper and I was telephoned by the Dean of Education, who asked me for a copy. Furthermore, and ironically within three years, an Army Adventure Training School *was* established with full-time staff, to design and manage courses to accredit trainers in all disciplines of outdoor activities. As a strong supporter, user and promoter of adventure training, I was proud to influence its introduction and popularity, and help the development of policies surrounding the purpose, planning and execution of adventure training as a genuine form of training to support the development of young leaders.

Queenscliff was a good place to live however; Lo and the kids loved it. Lo involved herself with a local children's charity and, as is her way, organised a children's concert. Held in the local town hall, it was a huge success, both with the children performing and the locals, who donated significantly to the coffers of the charity. The year at Staff College ended, an enjoyable year but just a tick in the box for further promotion. With great anticipation we packed for England.

24 POSTED TO LONDON

The Australian Army has a number of overseas exchange postings. The purpose of these postings is to develop closer links between our Army and other Armies across the world, in particular our close allies, the US and UK. Amongst these overseas postings in the 1980s, was 'Artillery Analyst, Technical Intelligence, Ministry of Defence, UK'. That was the job for which I had been selected.

I was to work in 'Defence Intelligence 60' (DI60), the UK's Technical Intelligence Branch. The branch comprised about 20 technical intelligence analysts, officers who were specialists in their corps, not specialists in intelligence. In addition to an artillery analyst, there were two armoured corps analysts (tank and cavalry), plus engineer, infantry, communications, missile and electronic warfare analysts. The analysts, at the rank of Major, were led by a Lieutenant Colonel and over us all was a Colonel. The analysts were experienced field officers who, using their practical knowledge of the use and application of 'western' weapon systems, would analyse the intelligence information available, and make assessments on the design, capability and characteristics of the 'eastern' weapon systems. This information would then be made available to military strategists, planners and weapon designers of the UK, after briefing relevant government agencies. This may include the Defence Secretary or the Minister, depending on the relevance of the information and how it might affect the status quo.

I had to be selected and informed of this posting more than a year in advance to allow the security clearances to be completed. I wasn't even fully briefed about my role before we left Australia, because that is the way the of 'intelligence' business. Information is compartmentalised and made available on a 'need to know' basis, so that if there is an intelligence leak, the damage will be minimised—no one person (except the very top of the tree, and perhaps not even then) is aware of everything. Until I had been comprehensively security cleared by Australia, the UK and the US, I could not be fully informed about the job.

However, I was told I was to be the UK's sole artillery analyst: advisor to the British Army and UK government on all matters pertaining to Russian and Warsaw Pact Artillery. In this job I was to gain complete and unfettered access to all 'UK Eyes Only' intelligence material pertaining to my job. In effect I was to become a fully accredited 'Pom'. It sounded a little ridiculous, a bit exciting, and more than a bit mysterious. What did I know about such things?

With three young children in tow, and an uncontrollable number of bags, we set off on our huge adventure. First stop: Singapore. It was completely different to when we left over 10 years before. We visited Ah Choo and her parents. They were safe and well, but something had changed. No longer were they living in the freedom of their own attap house in their kampong. They were now ensconced in a multi-story apartment block, in a characterless concrete building, the type that had sprung up all over Singapore. Next stop: Sri Lanka. Lo's parents were on one of their world trips and they met us at a resort called the 'Ranwelli', on the west coast of Sri Lanka. They were already there, and I remember Gwenn, Lo's mum, standing on the far side of a lagoon we had to cross by punt. She was wearing a white dress that looked stunning. Her poise reminded me of a colonial tea plantation mistress, not that I have ever met one.

Sri Lanka was such an amazing education for the children, who saw first-hand the different lifestyle, the beauty and the poverty, all mixed in together. They were a little young, but Rae does remember having Lo take her dress straight off her and handing it to some poor little girl. From Sri Lanka we flew to Rome, upgraded to first class by Air Lanka Airways. The little Fiat taxi at Rome's airport groaned as we piled in: Matt had to lie on his side between the top of the baggage and the roof. Our hotel room had no spare floor space, only beds, upon which we walked to get around. Matt very nearly mistook the purpose of the bidet. At last, we headed for London. All the way we had been flying business or first class, as was the way of the Defence Department for members posted overseas.

After a couple of weeks in an apartment in Oxford Street, we settled into our married quarter in Coldstream Gardens. Originally a 'married patch' for the officers and families of the Coldstream Guards Regiment, it had now been opened up to officers from a wider background. Coldstream Gardens was in Putney, on the Wimbledon Underground, only about 10 km from the centre of London. My daily commute from Putney to Embankment Station next to the Ministry of Defence (MOD) was about to begin.

We were well looked after by the Australian Government. The children went to 'preparatory schools'—very posh indeed. Heating and lighting were taken care off with only a minor contribution from us, which was just as well as the heating bills in England were an entirely different proposition to what we were used to in sunny Australia. Our three-storey, semi-detached house soon became known as the house that was always warm in winter, because we didn't need to concern ourselves about the cost.

Almost immediately, I felt a connection with my work. I loved the feeling of independence that we, as officers, were accorded in the British military. I thought it strange at the start that I, as an Australian, was the primary person responsible for advising the British MOD about the status and future development of Soviet artillery. But that was the way the Brits played it in this branch. And I wasn't the only foreign officer working in DI60:

there were two others, a Canadian and an American. The American, whose name was Gary Gee, was the guided missiles specialist, basically on loan from the Americans. He came from Alabama—a real southerner and a gentleman. He'd come to work about seven o'clock each morning and go home early, in the manner many Americans are used to. I was more in the British mould, getting to work around nine-ish. The Canadian was there as a small arms analyst.

And so, here I was, the Soviet and Warsaw Pact artillery 'specialist'. The Brits relied on me to know my stuff, but it was all new to me. I remember my predecessor, Major John Cox, saying to me on the last day before his return to Australia:

'Okay, Ben, this is it mate. You are now the UK resident expert on Soviet artillery!'

It gave me sleepless nights at first. But I swatted and worked hard and delved into the files that had been left for me. When something significant was discovered by any of us, we briefed firstly the Colonel, who might brief at higher levels, depending on the significance, or we might brief the Defence Industry scientific community, including the British arms manufacturers, or we might brief directly at the highest levels of government. The Brits were strong believers in allowing the analyst who discovered the intelligence information to personally brief the topic all the way to the top. This ensured that the information was not misrepresented or diluted in any way and that, if the senior people had questions, the answers would come straight from the person who knew most about the subject, the actual analyst. At one point later in the posting I found myself briefing the UK Minister of Defence.

In truth, we may have been analysts back in the comfort of the offices of the MOD, but this was the Cold War, and we were actually very much part of the front line, even if that wasn't in a fighting sense. This was the time of Reagan and Thatcher, the supreme right wingers of their respective conservative parties, who were hard-liners and regarded the Soviets as the 'Evil Empire'. Those were the days when US B52's left Fort Wainwright in

Alaska every day, loaded with bombs containing nuclear warheads, ready to attack targets in the Soviet Union on the word of the president. Nuclear war was just a couple of turnkeys away every minute of every day, at the whim of a US movie-star-turned-president. Coming from Australia, it was an environment that was new to me, but it was one that I quickly understood, and I attacked my work with a deep sense of responsibility.

We collected information on the Soviet Union and the Warsaw Pact countries, including Czechoslovakia, Hungary, Poland and East Germany. However in DI60 we were intelligence analysts, not spies. Anyway, no one spoke of spies, the language was 'sources' of information and the people who were part of those organisations were referred to as 'collectors'. We, the analysts, directed the collectors so that they understood our priorities. Information was collected covertly, through planned operations, or overtly, through open sources such as newspapers or military parades.

'Hands on' collectors included BRIXMIS, the British Commanders'-in-Chief Mission to the Soviet Forces in Germany. Quite a mouthful. BRIXMIS was an organisation equipped with fast patrol cars and manned by professional British soldiers, who roamed the countryside of East Germany looking to spy on Warsaw Pact military manoeuvres. BRIXMIS was an anachronism of the immediate post WWII period, when the Allies and the Soviets were rushing to capture Berlin. The Soviets made it first, but allowed the city to be carved up into sectors between themselves and the British, French and Americans. As part of the carve up and to enable the invading armies to 'keep an eye' on each other, missions were created by each country with special powers to roam the streets in the other sectors.

Other intelligence sources included human intelligence, referred to as HUMINT, from informants, defectors or asylum seekers, or simply 'open press', where information was released by military authorities, usually by mistake, to newspapers or television. Often these open press intelligence gains were the tiniest of snippets, but they helped us analysts tie together the bits and pieces we'd gleaned from all the other sources.

Secret and often detailed intelligence information was gleaned from satellites or aerial photography.

Valuable intelligence was gained from the physical exploitation of war materiel, such as rockets and projectiles, bits and pieces of artillery hardware, expended or 'live' ammunition, or anything that our collectors had somehow acquired such as a gun sight, or a fuse or a bit from a rocket tail.

The British Military Attachés stationed in Warsaw Pact countries were a goldmine for us. I soon realised that the accusations that were frequently aired in the press about attachés spying, that caused them to be 'shamed' and sent home in tit for tat expulsions, were mostly true. They did spy! Not overtly, but by keeping their ears and eyes open to glean information and accessing local sources like the press, or disaffected, dissident political groups or simply taking photographs of military barracks in the hope of snapping a picture of something interesting.

SIGINT or 'signals intelligence' was another huge source of information where radio, telephone and telex messages were intercepted and analysed by GCHQ, the Government Communications Headquarters at Cheltenham, a place made famous by the intelligence exploits of WWII. Another major source was satellite photography, acquired and analysed by JARIC, the Joint Air Reconnaissance Intelligence Centre. At each of these places I quickly got to know my specific contacts and would visit them frequently. Telephone calls might be intercepted, so there were many train and plane journeys.

This was all very new to an Aussie officer, more used to being with soldiers in the field than with intelligence collectors, scientists and politicians. Fortunately, the work was not all ultra-serious nation-saving stuff. There was also much fun to be had in working with Poms, who I found had a great sense of humour in the main. When John Cox was doing his handover to me, apart from discussing the job, the contacts and the systems, he had to brief me on the day-to-day mechanics of the job. Amongst other things, this included the combinations of the ten or so safes that I would be managing. No computers in those days, everything was on paper, locked behind steel doors. But what I found most amusing was his black attaché case with 'OHMS' (On Her Majesty's Service) written on the side in large gold letters. It was an old fashioned looking thing, the kind of strong leather case that you might imagine Sherlock Holmes to be carrying. The kind of case that attracted everyone's eye and said: 'look at me, I'm carrying something important, can't you see the OHMS on the side?' However, what really caught my eye were the many little shotgun pellet holes in the side, partly obliterating the letters. Very obvious!

'Can't the MOD afford to give you a new case, John?' I asked.

'Eh, the MOD doesn't know yet, mate!'

He realised the quizzical look on my face warranted a better answer than that.

'I was having a drink in the pub with the RARDE boys, the Royal Armaments Research and Development Establishment team. You've yet to meet them. Anyway, the bar keeper had an old pump gun mounted on a rack behind the bar.'

'What the hell is a pump gun?' I interrupted.

'A pump gun... a flare-muzzled shot gun mate, real old! The barman reckoned it still worked, but we were sceptical, so he offered to prove it. We'd had a few, so I challenged him to put a hole in my OHMS briefcase. And the bastard did! The bloody thing worked alright!' he said.

'So... where does that leave me, John? What am I supposed to tell the MOD?' I said.

'Nothing. Just use it and hand it over when you move on,' he said.

So I used it for my two years, drawing funny looks from commuters. I wondered what went through their minds. An antique OHMS briefcase with a multitude of shotgun holes tells a story that the commuters on the tube could only guess at. After my two years of service, I handed the bag over to my successor, Peter Overstead, in the same condition, and recounted the same story.

Within days of my arrival I was invited to have a beer with the mad scientists at RARDE, the boys from John's pump gun story. These were two technical specialists who worked at Fort Halstead, the home of RARDE at Sevenoaks, just to the south of London, a short train trip away. They had a reputation that preceded them. One was a huge, permanently grinning man, who was always happy and never seemed to wear a coat, even on the coldest of London days. The other was an archetypical 'mad professor', with curly black hair shooting out at all angles, black glasses and a skinny little frame, with a mischievous glint in his eye. I instantly liked them.

When I stepped off the London to Sevenoaks train, I saw these two odd-balls coming towards me.

'You must be the new Australian?' they said in unison. 'Come with us, you need to understand how we do things here.'

I felt welcomed and yet suspicious. There was something amiss about the way they greeted me, a bit like little kids, with a surprise just around the corner that they can't wait to show you. Sure enough, I was surprised! Just around the corner stood a jeep, the US WWII variety, with no top; and this, on a day when the snow lay at least half a metre deep on both sides of the road! They laughed heartily as they saw me gape: told me to sit on the back seat, which was even more exposed than the front seats, and set off, laughing all the way. Yep, it was freezing! They were so excited, sneaking looks over their shoulders to see if this new Aussie was up to it. I gathered that I was being tested and so I sat and shivered and smiled. What else could I do? Instead of being taken to the establishment of RARDE, we went straight to a local pub where they poured pints of beer into me for a few hours, then took me back to the train station where they shoved me into a seat.

'What about RARDE?' I asked.

'Next time' was the answer. 'This was more important! Now we know we can do business with you.' I wondered if this was how I was going to do business with my other so-called 'sources and contacts'.

I soon realised that us Aussies are just beginners when it comes to beer drinking. Well, bitter beer that is, at room temperature, which at first is abhorrent to most Australians, but in the end, I came to enjoy very much! Real beer as the Poms put it. Who wants to drink icy cold beer on an icy cold day? Better to drink a room temperature real ale! Beer drinking turned out to be a popular way of dealing with my various British-based compatriots, including in particular my communications intelligence desk officer at GCHQ. My first contact with him was also memorable. Again, a pattern was set. He was a huge man called Ian, who would always meet me at the train station at Cheltenham in his little Citroen 'ducky'. To see this man, who was about 6'6" tall with feet that looked like boats, somehow slide into the little car, and almost fill it all by himself, was a remarkable sight. It was Ian's job to read through the endless transcripts of communications intelligence, known as SIGINT, and identify any reference to artillery, in any form. The Russians, well aware that many of their phone calls and messages were intercepted and read by the intelligence community of the west, referred to many of their weapon developments by code names. These code names in themselves then became the target of analysts like Ian, who would scour the transcripts looking for clues. Once a known nickname was revealed, or a potential new one identified, he would translate the sentences that led up to and surrounded that nickname, to see if the discussion was useful. Computers that do that sort of thing today were not around, so it took a significant amount of time.

Ian was good at his job. He was also good at drinking beer—Guinness beer, by the pint, which seemed to flow straight to his oversized shoes, as he appeared entirely unaffected by the pints I saw slide down his throat. I received calls from Ian once every few weeks, requiring me to catch the train to Cheltenham. That in itself was an experience. 1st Class of course:

'Cup of tea and a toasted sandwich, Sir?'.

'Thank you steward, yes', whilst reading The Times. Very British, all this.

I quickly got used to it. However, the beer drinking always made the return journey by train somewhat risky. The temptation was to fall asleep.

I remember one occasion in particular, after a few pints more than usual at the lunch meeting, waking up on the train and wondering why it was that the train was now going in the opposite direction. With an aching head and somewhat slow brain, I studied the stations as the train sped on and it dawned on me that the train was heading out of London, not into London. I was heading back to Cheltenham! I'd fallen asleep and stayed asleep, right through the long stop at the terminus at Paddington, and not woken up until halfway back to where I had started from. That embarrassing drunk in a suit in the corner, fast asleep, that everyone ignored, was me! I quietly slid out of my seat at the next station and caught the next train back to Paddington.

I was told that I would have to deal with 'real' spies, the men from MI6. This was the same organisation that had spawned a thousand spy stories, including James Bond. MI6 was responsible for conducting clandestine missions outside Britain's borders, whilst MI5 was responsible inside the UK. My introduction was hilarious. None of the men I met would give me their real name, and the first one I was introduced to was dressed in a grey mac—an overcoat which somehow made him stand out and said: 'Here I am ... I'm a spy!' These were the men who would alert their various operatives all over the world to what the analysts in DI60 were after. For example, I might be in the process of confirming an analysis in relation to, say, the technical capability of a weapon system, such as the ignition process, detonating mechanism or aiming device. The collector would then be instructed to find out somehow, either by stealing a part of it, or buying a part of it, or acquiring the whole weapon on the black market.

More than once did I receive a call from an MI6 man in London about an operative in some place like Peshawar or Islamabad, asking for instructions on what to do next in relation to acquiring the said article. MI6 would want to know how important it was, how much risk was warranted and sometimes, how many US dollars should be paid for the article that was available on the black market. These calls would come at any time. Suddenly, perhaps on a Saturday morning whilst shopping with the family, or on the tube in the middle of the commuter crush, I would be asked to make an assessment on the spot, over the phone. My response would then be relayed all the way to the war zone in the Middle East or Pakistan or Afghanistan, where a 'real' British Secret Service Agent would act on my instructions, steal the item in question, or hand over thousands of US Dollars to a black market profiteer, who was more than likely some Mujahideen tribesman who had stolen the item from the Russians.

Then, usually a few weeks later, I would be informed that the item in question was on its way to Brize Norton, the RAF Airstrip to the west of London. My two RARDE compatriots and I would meet the aircraft, mostly in the dead of night, and take receipt of a package, or an entire rocket, which would then be carted back to RARDE for 'exploitation', sometimes in the back of their unmarked Jeep, covered by a tarp! We had an 'agreement', a 'special clearance', and there were never any customs officers present. It was illegal... but legal. No authority seemed to know how to deal with this stuff, so we were ignored. Once back at RARDE we would pour over the item like excited children and poke at it and X-Ray it, and postulate about it before the real detailed assessments were commenced. In these days of heightened anti-terror laws and restrictions, it is hard to imagine us belting along a British motorway in the dead of night, with a rocket still filled with explosives, hidden under a tarp in the back!

From conducting hands-on analysis, we gleaned a lot about Soviet engineers, their design philosophy, and the nature of their manufacturing processes. For example, we learned that the Russians were less than precise in the manufacturing of their weapon systems, certainly those that were classified as artillery delivery systems. Quite often we found, for example, that the welding on the inside of their artillery rockets was rough and poorly executed, almost erratic, with lumps of solder all over the place, as if the job was done in a hurry. This could make the rocket less predictable in its flight, affecting the trajectory and thus the accuracy. The fuses were often quite rudimentary and may not have worked. This was entirely unlike the quality control employed by a majority of western arms manufacturers, where perfection and precision were required to satisfy the demanding specification of contracts for the delivery of all weapon systems. However, this meant that the western cost of manufacture of an item was significantly greater than the Russian equivalent. We believed that the Russians cared less for precision and more for quantity. They simply made more. Their philosophy appeared to be that, if the rocket or the warhead or the bomb did not detonate as required, well never mind, they would just fire twice as many! After all, it cost less to make them.

Russian designers clearly didn't care much about ergonomics affecting the health of their soldiers either. Perhaps they thought they could simply supply more of those, too! I got my hands on a video made inside the turret of a 125mm Russian self-propelled artillery piece called the 2S3. It showed the gunner loading the gun by ramming a round into the breech and firing the gun, which is the normal thing he would do. However, the recoil of the breech and barrel and the resulting blast wave seemed to just about blow his head off. Nevertheless, he was filmed as he kept going, another round, and again, and each time his head jerked so violently you would think it was about to blow clear of his body! On a western weapons platform, there would have been a sophisticated system of reducing the effect on the gunner, in both noise and blast.

Being a part of the intelligence community was a bit like being in a club. The Brits would do technical intelligence exchanges with other countries and agencies. The amount of information that was exchanged, and the quality and source of that information, depended entirely on which country. The closest intelligence community in the world then, and now, is the AUSCANUKUS grouping, which refers to Australia, Canada, the United Kingdom and the United States. It is variously referred to as the 'Four Eyes community'. To this day, that remains a true statement. New Zealand was a bit on the nose in the eighties for not allowing US nuclear-powered ships to dock. These days they have been forgiven and the community has an 'NZ' tacked on the end, and the grouping is now called the 'Five Eyes' community. Within that group there would be special one-on-one relations, and if the country involved wanted to make a statement that might cause some political fallout, a term 'Eyes Only', preceded by the country, was attached. Curiously, although I was an Australian, I was not privy to 'Australian Eyes Only' but *was* privy to 'UK Eyes Only'. This was because I was wholly seconded to the UK and worked exclusively for the Brits, not the Australians. This was a level of trust that was placed in me as a person who the government of Australia had offered up to the government of Britain to work entirely and exclusively for them.

This relationship was unique and only understood between Australia and the UK, clearly demonstrated by the fact that every time I went to the US to visit the CIA at their headquarters in Langley on the outside of the Washington beltway, the security people on the CIA front desk would take the unusual precaution to lock me in a room for a couple of hours whilst they checked my credentials. They never seemed to remember the last time I was there. I was subjected to the same polite but firm interrogation every time. Invariably I would arrive at the giant entrance hall of the infamous CIA headquarters building, the one shown in thousands of movie and TV clips, and walk past the famous wall of stars with names of slain agents. Feeling very self-conscious, I would open my Australian mouth to declare who I was and who I was seeing, and every time their first reaction would be one of shock. Without fail I would be escorted to their little sound-proof room, the one they use to stow all hapless characters of dubious repute that deign to enter the hallowed halls of the CIA without what they consider to be the right credentials. There I would sit, in a sound and electronic signal-proof room behind a locked door, for an hour or more, until politely but unapologetically someone would fetch me!

Within the AUSCANUKUS group, the exchange of information was an ongoing thing. As UK analysts in DI60, we would visit our US counterparts at least twice per year, and they would also come to London. We came to know each other's families and became friends. As well as the CIA,

I would deal with other US agencies such as the National Security Agency (NSA- the signal intelligence people), the Defence Intelligence Agency (DIA), and the Foreign Science and Technology Centre or FSTC, in Virginia. Located in Charlottesville, FSTC was in the heart of the south, the heart of the old Confederate states, where life-sized statues of Generals Robert E. Lee and Stonewall Jackson adorn public parks and where, only a year before I was there in 1985, the Klu Klux Klan had burned a cross in the front yard of the African-American boss of my intelligence counterparts in FSTC. Here I was, chatting intelligence matters with American guys who seemed to belong in an American Civil War movie.

I happened to be in the US the day the Chernobyl nuclear reactor imploded. Sitting with my US counterparts in FSTC, we were suddenly interrupted by an excited agent! 'Come and see this', he was saying, actually addressing the US guys, not me. But in the excitement of it all I was carted along and found myself standing in a large room with screens the size of the wall looking straight down the barrel of the nuclear disaster that was unfolding in Russia. Within moments of the explosion, the Americans had diverted one of their satellites to cover Chernobyl to study the disaster first hand. This lasted for a few minutes until someone in the room looked at me and asked what this unauthorised bloke was doing here, so I was shuffled out. But it was very interesting while it lasted and gave me an insight to the capability of US satellites.

I did enjoy going over to the states. Driving some big American hire car was always fun, and the east coast of America is very pretty in parts, particularly the Appalachian Mountains; Johnny Appleseed country, where it was entirely possible to see a bear in the woods just outside of your country diner or motel. After all, Northern Georgia was where 'Deliverance' was filmed, a legendary movie with legendary scenery, depicting people that are not hard to imagine as one drives around those parts.

I took Lo on one of those US visits. It was a memorable trip that very nearly didn't come off. I was hoping to get her onto the RAF flight that went to Washington once a week, but when we got to the airport at RAF Brize Norton, the RAF clerk decided there was no room for her. I'm sure he thought that this bloody Australian is 'using the system'. Lo's parents were with us in London so were able to look after the kids, and they had driven us to the airport. However, the RAF Movements staff claimed the flight was full, so the only alternative was for Lo's parents to take her to Gatwick and try to get a commercial flight. We would then have to meet up in Washington. But as we tried to untangle our luggage, which had been piled into the one suitcase, and Lo started to unravel her underwear from mine and pack it into a plastic bag, the RAF Airman behind the counter turned slightly red and took pity on us poor struggling Australians. After consulting his boss— obviously to tell him how these undignified Aussies clearly couldn't afford two suitcases—he let Lo on the aircraft. Lo's underwear was stuffed back into my bag and she won the day!

The RAF Comet that was to transport us had its seats facing backwards: very disconcerting! But, it was free! For my sins, I was nominated as the flight officer-in-charge, second only to the pilot in authority. How ridiculous I thought. I soon realised why none of the English officers wanted this honour. Throughout the trip, pompous English officers complained to me about absolutely everything: meals, lateness of take-off by a few minutes, poor service; anything that upset their precious little world. I suddenly became aware that not all Pommy officers were of the type that I had come to know in DI60. I was not expecting this display of outright bad attitude. I nodded wisely to every complaint, writing nothing down. When the pilot asked me at the other end was everything alright? I answered that it was a just fine, an excellent flight, no complaints.

Lo did get into a small panic when we nearly missed the end of the runway due to the thick fog over Washington. Just as the pilot was telling us all was well and we would soon be landing, the aircraft burst out of the low cloud, only to find it was about to land on the grass approach to the runway. The piano keys were not in sight! We did think it strange to look out of the window at pastures and trees, and the way the pilot gunned the engines was definitely not normal. But the pilot re-assured us all, and the second attempt went well. I didn't receive any complaints about the aborted landing—the pompous Poms were all still frozen to their seats, scared shitless!

It was a memorable trip. One of the analysts in London, knowing that I was taking my wife, had arranged for me to see the rocket specialists in Huntsville, Alabama. This meant an 800-mile journey down to the southern states—an excellent chance for Lo and I to see the US 'deep south' in our little hire car. We had changed the big Yank tank that we were first allocated for a VW Rabbit, the US version of the Golf. Unfortunately, the Golf broke down right near a town called Chattanooga (of the Chattanooga Choo Choo song fame), forcing us into a dodgy little motel for a night, where the sheets still had evidence of someone else's pubic hairs. Not very savoury, but an experience. It was around there, somewhere in the Appalachian mountains, that we ran into a very friendly couple in a classic US diner, the type you only see in the movies. They chatted about this and that, but had no idea where in the world Australia was, and when I explained that Australia is in the southern hemisphere, where it gets hot in the north and cold in the south, opposite to what happens in America, the wife was so stunned that her husband patted her on the hand and promised he would explain it to her later.

On that same trip, I had been nominated by JARIC, the UK satellite imagery organisation, to accompany the weekly boxes of satellite imagery back to the UK. These were highly classified—TOP SECRET 'Codeword'— and were carried physically across the Atlantic on these RAF aircraft. The RAF aircraft were still subject to customs, but not the boxes with the satellite imagery, nor the official escort, in this case me! Lo was already aboard the aircraft, but I had been asked by airport officials to stand with them on the apron, at the rear, near the tail. Unsure of what was happening, I was suddenly surprised by a row of black limos which drove up in a hurry, seemingly out of nowhere. Out jumped a bunch of armed agents in dark suits, who took up positions, providing a 360-degree cover. For a moment,

I thought I was acting out a movie... black limos, secret service agents in identical clothing with dark sunglasses, looking every bit the part. I just stood there until I was approached by a man asking me to count and sign for these boxes, 21 of them. I signed. But, we had done some shopping, and I had a few of my own boxes to add to the list. No-one seemed to mind, which surprised me. That was the eighties! It would never happen in the current world of terrorist fear!

Proceedings at the other end of the journey were just as bizarre. I was met by a smart, well-dressed Sergeant in his finest regimental uniform, who got such a shock when I opened my mouth that he refused to believe that I was the designated escort. Clearly, he had not been briefed. I was used to this treatment in the US, but not here in the UK. He had to ring his boss, who also didn't know about me and wanted to speak to me. Here I was, just wanting to get home, and these Poms refused to believe I, an Aussie, worked for them. More phone calls. Finally, they were persuaded. Then the next problem. He had brought along a truck that was too small for the 21 boxes plus my three boxes. Lo had long since gone off home to Coldstream Gardens. So, to solve the room dilemma, I suggested he stay put whilst I escorted the imagery boxes by myself. The poor man nearly had a heart attack. No way was he going to let this Top-Secret cargo out of his sight in the care of some dodgy Australian! His only choice was to lie on top of the boxes between the roof of the van and the top of the boxes, in his regimental best. It reminded me of Matt in the taxi in Rome. But, we got there. The Sergeant was happy that he was able to prevent an international imagery snatch by the Australians and I was at last able to catch the late train home, with my three boxes. Another story for the boys, I thought.

When the intelligence exchanges were conducted with countries outside of the AUSCANUKUS group, the protocols and processes were more formal and deliberate in their planning and execution. The intelligence sharing relationship with NATO was always guarded, because it was generally accepted that some of the NATO countries 'leaked'. Nobody pointed any fingers, but the politics in some of the European partners meant care had to be taken as to what information was shared.

We did an exchange with the Bundesnachrichtendienst, or BND, which is the German Federal Intelligence Service, that was quite un-German-like, notably for the amount of alcohol they poured down our throats, rather than the intelligence goodies that were shared. With everybody loaded into a large tourist bus, they took us into the woods outside Dusseldorf and drove us to a large, walled compound which enclosed a beautiful old stein. The location seemed a classic 'secret' hideaway: a walled castle in a dark forest. We proceeded to discuss technical intelligence issues whilst drinking litres of German beer, which is not known to be low in its alcohol content, with each litre backed up by a schnapps chaser. We hadn't yet had lunch! I recall that the German analyst I was talking to was more interested in asking me about Australia than talking about Russian guns.

We ate and drank most of the day, and swapped notes about the bad guys we were supposed to be spying on when, late in the afternoon, they put us back on the bus. We, the team from DI60, were all heading to Berlin to discuss collection requirements with the BRIXMIS boys. This was the first and only time, however, that I was treated differently than the Brits on our team. The Australian government forbade any member of the Australian military from setting one foot inside communist East Germany. In those days, the East Germans didn't allow any road traffic from the west—this was the Cold War, and as Berlin was surrounded by East Germany, there were only two ways in, by train, or by aeroplane. Tensions were always high. Since I was not allowed on the train, the plan was to drop me off at a railway station where I could catch a train to the airport and fly to Berlin, while the rest of the team went to the 'Frankfurt Bahnhof', where they would catch their train to Berlin.

We had all consumed way too much alcohol and we lolled about the bus in ignorant bliss, half asleep. Then suddenly the bus stopped.

'Lans!'

'What?'

'Out! This is your stop!'

'Oh, right then,' as I struggled to remember why I was being thrown off the bus.

I clambered out and promptly disappeared, slipping straight into the gutter.

'Lans, where the hell are you?', a few concerned voices asked. 'There he is... he's okay! Let's go!'

I looked up to see a couple of cheery faces smiling at me through the rapidly departing bus windows.

I was actually lying in the gutter, in my suit, looking up and waving goodbye and no-one seemed to think that was strange, neither me nor my English mates on the bus. Well, we didn't, but the German onlookers were probably amazed and more than a little judgemental. Stupid tourists! Little did they know! I didn't care.

I got myself up and struggled into the station, managed to buy a ticket, even managed to get the right train, and sat down trying my best to recover my wits. By the time I reached the airport I was nervous that the efficient German officials would give me a hard time again. On the way into Germany I had been stopped by the one of the Airport Immigration Officials, who looked at my passport, looked at my travel order, which was a NATO travel order, looked me in the face and calmly said:

'Australia is not part of NATO, why are you travelling on a NATO Travel Order? Step over here please.'

'Ah, well you see, Sir...' and there it began again. 'I work for the Britishers you know...' (Remember those blokes that beat you in the last war?). Luckily I didn't actually say that or I might have been deported back to Australia, never mind the NATO Travel Order!

I talked my way through that time. These uber-efficient Germans! No other border crossing officials had ever spotted the obvious. Trust the Germans to be the only ones.

Nearby to Berlin was a large Russian and East German training area and live firing range called Letzlinger Heide. The BRIXMIS agents would attempt to covertly intercept convoys of military vehicles and take photos of them. The photos were useful to help identify new self-propelled guns, or tanks, or any military vehicle and help identify such things as upgraded sighting systems, electronic improvements (by spotting new antennas or other electronic gadgets) or add-on armour and other modifications. The Russians and East Germans were well aware that BRIXMIS may be out and about trying to photograph them, and as a result, they almost always covered their vehicles and weapon systems with large tarpaulins. This forced us to become expert 'tarpaulogists'. We defined a tarpaulogist as... 'one who studies the lumps, bumps and assorted shapes under a tarpaulin in a desperate effort to determine what is underneath'. The typical add-on armour that is seen on many tanks and armoured vehicles these days, in the east as well as the west, was first invented by the Russians and first discovered by the BRIXMIS team, by taking photos of the convoys entering and leaving Letzlinger Heide.

My short stay in Berlin etched itself into my memory forever. Berlin was divided into four sectors. There was the Soviet Sector: it comprised about half the city. The other half of the city was divided into the US, French and British Sectors. I only had authority to enter the British Sector. The BRIXMIS lads were waiting for me and took me to the British officers mess which was situated right next to the 1936 Olympic Stadium. This mess had once been Hitler's own mess, a stark, colourless brick building. It was not the only time that I was to come across facilities used by the British that were once part of the German establishment of the Nazi era. As soon as possible I left the mess to check out the Olympic Stadium. This large mausoleum-like stadium was something else, it gave me goose bumps. I had heard so much about it. This is the place where my own father sat in 1936 and watched Jesse Owens, the famous US athlete, defeat the cream of the golden-haired German youth, much to Hitler's disgust. I can still see my father's face as he told me the story.

My father imported Indonesian spices and coffee and tea into Holland and Germany. He was on a business trip to Berlin and decided to take a detour to visit the Olympic Games. This was where Hitler showed off his country's ability to manage and organise an event, the likes of which the world had not yet seen, with the precision and attention to detail that was to mark the Third Reich in the years to come. As I stood at the entrance and gazed firstly into the stadium, then up at the empty stands, the hairs on my arm stood up. I could hear the cheers, see the fanatical crowd, and picture my father here somewhere, in this cold, eerily grand place.

At the entrance to the stadium stands an imposing bell tower. Later, back at the mess, I was told that the bell tower has a special significance. Because of the division between east and west Berlin, created by the Berlin Wall, many triumphs—and sometimes tragedies—were played out. East Germans frequently attempted escape to the west, either to join family members from whom they had been cut off, or simply to find a better life. These attempts, early in the days after the wall was first constructed, were often as simple as running at the wire and the wall, with the hope of climbing or jumping over, or even crawling through in the hope that guards would not shoot, or would miss. But later, as the wall became more secure and well-guarded, it necessitated complex, well-planned escapes, including tunnels and gliders and air balloons. Over 1000 people were killed trying to get over or under the wall: between 1961 when the wall was built, and 1989 when it was torn down, more than 5,000 East Germans including 600 border guards, managed to cross the wall from east to west, to freedom.

There were escape attempts nearly every day, and I was told that in the event of a successful attempt, the bells in the tower would toll, ringing a fast, joyous chorus of bells across Berlin. In the event of a failed attempt, where someone had lost their life, the bells would toll slowly, in respect, as a sombre reminder of the cruelty of that wall. On my second night in the mess, I heard the bell toll... slowly, and I wondered about the person who had just lost their life, probably in an attempt to re-join their wife, husband or sweetheart.

The next day, before the rest of the D160 team arrived, I caught a tram to the Friedrichstrasse for a walk in historic central Berlin. It appeared to be a grey, dour city, and the West Germans that I met had a remote, even detached manner about them, in apparent defiance of the tragedy of their torn and divided city. I walked to the infamous 'Checkpoint Charlie', one of only three crossing points between the Soviet and Allied Sectors of Berlin. It was bitterly cold, I doubt if I had ever felt colder in my life, but as I walked along the Berlin Wall, the cold seemed insignificant as I became entirely engrossed with the ghosts of history.

The wall was a vibrant reminder of the cruelty of man, which was depicted in the graffiti on the wall here, right in front of my eyes. Very graphic! People had painted poignant images on the wall or written the names of loved ones they could no longer see because they were now on the other side of the wall. There were poems of loss, of love, of hopelessness. There were names of escapees who had died in just that spot. The graffiti expressed cries of desperation at the folly of such a wall.

A small museum at Checkpoint Charlie showed images of some of the dramatic escapes over the years, such as the picture of the East German Guard looking back to the east with frightened eyes as he put down his weapon to step over the temporary barricades and escape to the west on the first day of wall's existence. There was a picture of another guard actually holding up the barbed wire to release an escapee into the west, and the picture of the man passing a baby through a hole in the entanglement to another person, presumably his wife, on the western side. It seemed the weight of the world, the very threat of the cold war, and every symptom of humanity's propensity to wage war, resided right here, right at this point of the wall. This ugly wall represented all the world's evil and cried out with sadness. I felt so insignificant, so small and helpless, but I couldn't tear myself away.

I climbed up onto the viewing platform from where I could stare into the east, at the infamous Brandenburg Gate with the charging horses on top; then I walked some more, through the nearby forest where it is said that Hitler's bunker was located, the one in which he committed suicide. I was not to know that in 1989, only four years later, the wall would be torn down, brick by brick, through people power, marking the end of the Cold War. I was in Australia at the time. We were living in Braidwood when I received a message to come in to the Defence Intelligence Organisation in Canberra. There, on the table, lay a parcel for me. It contained some lumps of concrete with bits of paint on them, and a few photos. The parcel had been sent to me by the team of BRIXMIS. With the breaking down of the wall, the requirement for the existence of the BRIXMIS team was no more. When they were disbanded, they souvenired chunks of the wall and sent them all over the world to intelligence analysts with whom they had worked. I received such a parcel. It brought back a lot of memories.

25 THE UK MINISTRY OF DEFENCE

In my time as the principal UK Soviet artillery technical analyst, I was fortunate to be the lead, or play a major part, in a number of significant intelligence assessments and findings that, in some tiny, miniscule part, helped to change the status quo of the Cold War.

Not long after I arrived at my new job in DI60, my predecessor John Cox, had an interesting item of intelligence to hand over. For some time, it was suspected that the Soviets were developing a super-sized Multiple Rocket Launcher (MRL). The calibre was thought to be around 220-240mm per barrel—such a rocket launcher would deliver a huge payload. John had got his hands on some photographs of what appeared to be a large rocket launcher covered in tarpaulins, being driven along a street somewhere in the Eastern Bloc. He estimated that the covered tubes were very large, perhaps as large as 240mm. This was big, as the average size of rocket launchers in the west was around 120mm.

The job of a technical intelligence analyst like me was to follow through on these theories and turn them into real intelligence by using as many sources of information as possible to confirm and build the picture. We had received some unreliable pieces of HUMINT that there was a new, large rocket launcher in the Soviet inventory, larger than the ones we knew about. At the same time, my man at GCHQ had picked up bits and pieces of information of an MRL being tested at a certain weapons testing range deep in Russia. This was unusual, because there was no reason to be testing the already fielded MRLs there. I alerted our people at JARIC, who studied the satellite imagery on a daily basis, to look for anything unusual at that test range. The problem was that the Ruskies seemed to be alerted to the scheduling of our satellites. There was little doubt that testing was going on, but every time the satellite took pictures, the weapon systems they were testing were covered under tarpaulins, until one day we had a break through! When the satellite passed, one of the vehicles was only partly covered by a tarp, and the analysts at JARIC spotted what could be an MRL with five large tubes.

A further breakthrough came when it snowed on the same range. The JARIC analysts knew the range like the back of their hands and could identify every new track or mark in the snow. The satellite had passed over a few moments after an MRL had been fired and returned to the covered sheds, and they were rewarded with pictures of a large burn mark on the snow, right at the firing point. With this information in hand, and the previous bits of intelligence, I enlisted the help of the scientists at RARDE and they were able to measure the burn mark, and subsequently determine the amount of propellant that would have made a mark as big as this. From that, an assessment was made that the MRL was likely to be big, real big, around 220mm to 240mm calibre possibly.

A rocket of that size would have a range commensurate with the propellant that it could carry, and a warhead capability far exceeding anything that the west fielded. Working with a variety of artillery, rocket propulsion and ammunition experts in RARDE, we were able to postulate the range and strike power of this weapon system. As this information was new, I worked with my counterparts in the US to seek their agreement to our UK assessment, and see if they had any additional information to offer. Then it was time to brief the Minister of Defence, which I did personally.

Finding the 2S7 was another exciting chapter. This gun was a 203mm artillery piece, self-propelled and with new capabilities. Once again, we had some warning through SIGINT that there was a large new gun around named the 2S7. The soviets have a nomenclature system for their self-propelled artillery that is quite simple. The 2S1 is a 122mm system, the 2S5 a 152mm system, the 2S9 is a 120mm system, and the 2S12 a 240mm mortar system, but the 2S7 was unknown. It was the only name mentioned that did not have a calibre we could link it with. As often happens, without warning, circumstance presents an opportunity from an unexpected quarter.

The British military attaché to Czechoslovakia, whilst driving through the country and passing a military base, snapped a photograph, as all the western attachés in the Warsaw Pact countries do when they see something that may be of interest to us analysts. He noticed that he had snapped a picture of the end of a barrel just as it was coming around a corner, and sent me a copy of his photo. John Cox had shown me images of other vehicles that were covered by tarps at various times, showing the possibility of a large barrel hidden underneath, and I suspected that these were potentially the new 2S7. But I was unable to verify the calibre.

Artist's impression of the 203mm calibre new Soviet Self Propelled gun referred to by the Soviets as the 2S7.
Source UNCLASSIFIED UK Ministry of Defence

What interested me about the attaché's photograph was that the end of the barrel and some parts of the carriage were clearly shown. My keenly honed skills as a 'tarpaulogist' came in handy and I was able to compare the details of visible bits in the photographs, with the shapes of the tarped vehicles that we suspected may have been 2S7s. The bits of chassis showing appeared to closely resemble a tank chassis with wheels and road track similar to the T80 tank, but the flatness of the platform suggested that it was based on what we called the 'Kharkov Engineer' chassis. John Cox, my predecessor had already alluded to the fact that the gun might be based on such a chassis. These possibilities then led me to compare possible overall weights. The capability of the chassis, if it was the one we suspected, with the combination of the wheels and tracks, suggested a weight distribution and a 'pounds to the square' inch ratio that would support a mighty superstructure and barrel.

I compared the bumps and bulges of what we knew from photographs and from the artist's impression drawn up by our branch technical staff under the direction of John Cox, and began to formulate the shape of the vehicle as a whole, with its barrel mounted. It confirmed a self-propelled system that did not have the usual turret, but instead had an open mount with the crew exposed.

I couldn't be sure of the calibre. However, after looking at the attaché's picture again and again, it dawned on me that there was a triangular 'Give Way' sign right there on that corner of the building around which the barrel poked. The distance from the sign to the barrel couldn't have been more than a few metres, so it would not be hard for our technical analysts to interpolate between the size of the sign and the calibre of the barrel.

This was an exciting development. Ah! But not so fast! When I rang the attaché with the simple question: could he go and measure a 'Give Way' sign somewhere please, so we can use it as a basis for measurement, his answer was:

'Ben, there are two sizes of Give Way signs in this country and I cannot remember which one that was!'

'Okay, can you go and measure that particular sign?' I asked.

I remember the long silence. Then he answered:

'Okay, I'll see what I can do. This had better be important!'

The next time I spoke with him some months later he was not quite as calm.

The 2S7, as it appeared in an open press Soviet publication three years after the gun was first suspected to be in production. This image verified the artist's impression of some years before, as prepared by the UK Ministry of Defence Technical Intelligence Branch.
Source UNCLASSIFIED UK Ministry of Defence

'Do you know that you nearly got me killed?' he said.

I was stunned into silence. He told me what happened: he had to wait before entering that part of the country again, lest he alerted the authorities. When he thought the time right, he and his driver returned to the area where he told his driver to park in a tree line not far from the barracks. He told him to stay with the vehicle, keep the motor running and doors open and be ready to bolt! It was evening, not quite dark. He needed at the very least, the light of the setting sun to see the sign. In a scene reminiscent of a spy movie, he crept to the edge of the barracks and, not seeing anyone, walked boldly to the sign with his measuring tape ready.

His story, already captivating, became more exciting. He took the measurements, but he was spotted!

'Halt!', came the cry in Czechoslovakian!

He could not afford to be caught, so he ran... the sound of a rifle being cocked in his ears! With bullets whistling around his head, he dived for the car, slammed the door and his driver raced him away. That was a close call. We had a few whiskeys and a few laughs over the story, several times over! Still, this was what our collectors did. They all knew the dangers. The information he gave me turned out to be very useful and we determined that the barrel was indeed 203mm.

This was a newly fielded weapon system, one that was unknown to us. New management practices and tactics had to be developed to counter the tactical and strategic effects of this weapon. Very soon, I was briefing the gun and its assumed capabilities to British forces stationed in the BAOR, the British Army of the Rhine, and to various allied intelligence organisations. The allied forces now had an additional heavy artillery piece to deal with. We worked on the potential maximum range of the gun. The SIGINT people at Cheltenham and other collectors then began to look for information to help me determine details of the natures of ammunition it fired and any hints as to its accuracy, not to mention the numbers that the Russians were manufacturing.

Within a few months we had determined that not only was this a new system, it was in high production and apparently was going to be distributed to all the Warsaw Pact countries. Most disturbing of all—it was nuclear capable! A Russian gun capable of firing tactical nuclear weapons! This gun became one of my principal briefing items whenever I was requested to provide an update on Soviet Artillery. Its superior range and nuclear capability gave it a tactical battlefield advantage that would require new counter-battery tactics to be devised, including how to locate such a weapon and understand its firing signature. However, as with the 240mm MRL, my part of the intelligence picture, although never complete, was the catalyst for other parts of the intelligence community. Slowly a comprehensive picture would be developed. DI60 was an inspirational place to work. Rarely in my career had I gone to work every day with such anticipation.

A defector from Czechoslovakia had turned up in Munich, asking for asylum, and he was offering some tantalising information. He claimed to be the designer of the new Czech gun called the 'Dana'. The Dana was a self-propelled 152mm gun developed entirely outside of the Soviet Union and was appearing in large numbers throughout the Warsaw Pact. I knew what the gun looked like, but knew little of its full capability, its range or fire power, rate of fire, reliability and whether the gun had a reason to be a threat or was just another artillery piece. I was asked to lead the interrogation team. Another analyst, who was not an artillery specialist but was there to help determine whether or not the defector was genuine, and a translator, were also present. For a week, we interrogated this poor man. It took several days just to gain his confidence and assure him we meant no harm. It was blindingly clear that he was petrified of authority. I obviously represented that threat to him, simply because I was the man asking all the questions.

Coming from a country where the military represented supreme authority, he had developed a mistrust of anything military. I did not wear a uniform, and I still don't know how it was he found out that I was a serving officer, but he knew. He refused to eyeball me, perspired profusely all the time, and literally shook in fear when he thought I was unhappy with an answer to one of my questions. All this went through an interpreter, and I had no idea how incredibly tiring such a process would be. Each day I would return to the hotel exhausted. However, eventually I began to gain his confidence. He wanted to go to the US. This was not normal, as he had defected to the British and should he be granted asylum, it would be to the UK. However, my boss back in London, Colonel Charles, made a few phone calls to our sister organisations in the US, and with no promises given, he was told it was not impossible, provided he was genuine.

How much of a role had he in the design of that equipment?

This is the fourth day I am here in this stuffy little room. It's hot in Munich! Why is it that interrogation rooms always look like interrogations rooms? Well, the way they look in movies, anyway. What would I know? We were supposed to be putting this bloke at ease, but this plain room with its bare, wooden floors, single table and stiff upright chairs is not exactly what I would have chosen to achieve this.

Every day this man opposite me sits there and sweats and refuses to look at me and we carry on this stilted conversation though the interpreter. I notice he has a brand new pair of Levi's on, and he keeps looking at my watch. Were the Levi's a bribe? Perhaps he wants another bribe? I'm still not convinced that he's an engineer let alone that he was the chief designer for the Dana. Maybe he was just a factory worker who knows a bit about the gun.

Ah, something is happening. The interpreter tells me that he wants a bit of paper and pencil. Okay, give it to him... let him go for it. I sit and watch him draw lines on a piece of A3. Hmmm, could be interesting. He has the hand of a draftsman that's for sure. I watch him for a while as the page fills with some sort of engineering diagram. What will this be? An hour passes. Now I'm the one starting to perspire. Why is it so hot in Munich? That's the trouble with these bloody European countries. Half of these bloody old buildings don't have air conditioning. Well that's what I reckon, anyway.

Okay, what's he showing me? Is this the inside of the turret? I think it is. I actually know what that looks like from the photos that the MI6 guys managed to acquire, but what he's drawing is much more detailed that any photo taken in half light. Now he's onto something else. Another part of the turret but in more detail. Hey, this bloke is good! He seems to have some real knowledge.

I tell the interpreter to keep him drawing for a couple of days. He tells me that in a couple of days the Czech man reckons he will only be able to draw the rudimentary drawings and to detail the entire engineering drawings of the gun will take months. Right! That's okay, I tell him. Just do the first level drawings and I'll take them back to London and get our people back there to look at them to see if he is genuine and we'll go from there.

This man was indeed an engineer. It didn't take long to prove his authenticity. For him to have that level of detail to reproduce by hand from memory meant that he had at the very least been closely involved with the design of the gun, even if he wasn't the actual designer. His drawings rolled out over the months, all to my desk and from there to the various organisations that were keenly interested in knowing just how the Dana systems worked. Within the intelligence community I was accredited with bringing this new system to the attention of the west, even though my part was simply to gain the confidence of a very vulnerable man who, it turned out, was indeed one of the designing engineers. It was an eye-opening experience to be involved in such an intelligence operation, but more significantly I was happy for the Czechoslovakian engineer, who I hoped was going to have a happy life in the US.

The Czechoslovakian 152mm Dana Self Propelled, automatic loading artillery system.
Source UNCLASSIFIED UK Ministry of Defence

Whilst in Munich, the other British analyst and I decided to celebrate by catching a train to the country, to a lakeside resort called Starnberg on the Starnberger See. We knew nothing about it, but the pictures in the hotel travel guide looked nice, so we thought we would give it a go. We took ourselves to the nearest bahnhof and looked for a train to catch. This was not too hard, until we tried to buy tickets. The system was fully automated: the instructions easy to read, if you could read German, but that wasn't the only problem: we had no change. What the hell? Who's going to care? Let's just get on and take a chance?

Well, we quickly found out who would care. The Germans care! Apparently, all of them. It seemed to me that Germans in general are law-abiding citizens, even to the extent that they only cross the street when the pedestrian lights are green. They do as they are told! Unlike us 'unruly' Aussies and 'arrogant' Poms. As we sat there, chatting and enjoying the beautiful countryside, who else but a ticket inspector came along? The classic 'fat inspector' of comic book fame, dressed in the finest railwayman uniform that looked vaguely like a WWII Waffen SS uniform, came striding down the aisle, looking every bit the man in charge. We could tell that this man was not a joke. We'd need to employ our best foreign diplomacy on him. Still in our suits from the morning's final interviews, we shrugged our shoulders and smiled helplessly at him, trying to explain:

'No change, you know? Sorry, could not get a ticket!'

Whammo! We were suddenly confronted by a fantastic interpretation of a NAZI concentration camp guard! His face red with anger and his hands shaking in fury, he demanded our passports. Oh, we thought, hmmm, he's not pleased! We gingerly gave him our passports and then realised that all the other passengers were looking at us with disapproving looks. This was obviously an 'un-German' thing to do. Bad Britishers! But what could he do really? Throw us of the train? Well... yes, actually! That's exactly what he did! He pulled the cord to make sure the driver knew to stop at the next station, walked us to the doors and, when the doors opened, threw our passports out onto the platform, looked at us angrily and pointed, AUS! AUS! Okay, we got the picture, we'll go peacefully! Geez? I felt like clicking my heels and giving him a 'Heil Hitler'! Luckily, I didn't. There was not a lot of humour to be seen in all this, for us anyway, and especially not the locals, as attested by the disapproving German faces peering out of the train windows.

Sheepishly we picked up our passports. True criminals, no doubt. There we stood as the train pulled out. Where were we? There is nothing around here. No train has stopped here since they lost the war, it seemed! Now what? We still had no change to buy tickets and what's more this station had no ticket machine, or ticket window at all. We looked around at the beautiful countryside in silence, almost expecting Julie Andrews to skip around the corner with her bunch of children in tow, singing about hills that are alive, but nothing happened. Bugger!

We moped about for a while until another train came, one that looked to be much slower and the type that actually stopped at all these little wayside stations, and there was nothing to do but to get on it. There we sat, shiftily looking about like naughty schoolboys, waiting for another fat inspector to appear and getting ready to make a run for it. But no one came and we got off safely at Starnberg. In the end this was a huge disappointment because we had failed to understand from the tourist booklet, that this was not tourist season, yet! Bugger, again! We found a bar that was open and sat outdoors in a sea of empty tables, had a few outrageously expensive beers, and decided to run the gauntlet back to Munich to our hotel once more. There were no further incidents on the train ride back to the hotel.

A few days later we were both back in London. Me, with a large roll of drawings under my arm and a good deal richer in life's experience. How lucky are we to live in a democracy, free to make decisions about all sorts of things about our lives, unlike the poor man I just interrogated for days and days.

One of the principal elements of this posting that made its mark on me was the sense of place and occasion that seemed to accompany everything that we did in the MOD. One time, when Colonel Charles took a bunch of us analysts to Paris to do an exchange with the 'Direction Generale De La Securite Exterieure', the French External Documentation and Counter Espionage Service. The French went to great lengths to make us feel welcome and special. The meetings were conducted at the 'Ecole du Militaire', the French Military Staff College, which is situated quite literally in the shadow of the Eiffel Tower, only metres away from the throng of tourists milling about just outside of the historic walls of the college. Ironically, the head of the French delegation was a man called Colonel Rose, a very English name, whereas, the head of our delegation was our boss, Colonel Delamain, a very

French name. Being the only English-speaking person in the room who could not speak French, I was allocated my own personal interpreter, a French Major who accompanied me everywhere. We had lunch in Napoleon's dining room. I was deeply affected by a sense of occasion; of being in the presence of the spirit, at least, of this great general, who had changed the fortunes of France and indeed, changed the politics of the new world.

Working in London in a headquarters in the centre of a large city was new to me and I missed the freedom of being able to go and play sport at lunch time. I soon sussed out a five-kilometre circuit from the Metropole Building, not far from Trafalgar Square, across Hungerford Bridge, down along the Thames to Westminster Bridge, then back along the Thames to the main building of the MOD, where I could shower and change. When I first arrived, I tried to use the single shower that was available in the Metropole building. However, I discovered that I had to share that shower with the cleaner. Not because the cleaner wanted to shower there, but because the cleaner kept all his mops and buckets in the cubicle, which I had to move out every time I showered. Running was a necessity for me as there was a constant temptation to eat and drink in this job. It was surprising, but I found that the English drink considerably more than Aussies, certainly during a working day. Every day, many a suited civil servant could be seen having a pint in the Sherlock Holmes pub across the road, or in the many small wine bars that dotted the 'arches', those small and sometimes large enclosed archways that are underneath the railways; the same arches that once upon a time housed the prostitutes and the 'down and outs' so prevalent in the stories of Sherlock Holmes, Jack the Ripper and the like.

On a Friday afternoon, we would sometimes knock off a bit early to have a drink on the 'Belgrado'. This was an old river steamer that was tied up permanently on the Thames just near the Hungerford Bridge. The ship was just a large pub really, with deck and internal bars, providing a favourite place to stand in the London sun and drink English ales with work mates. The real name of the ship was the 'Tattersall's Castle', but it had been dubbed the 'Belgrado' during the Falklands War, in honour of the sinking of the

Argentinian ship of that name. I was always amazed that I would be the only one to take off my coat and tie on a hot day, whereas all the Poms would leave all their coats on. For them, drinking in an open-necked shirt was just not on! One day, someone must have had a birthday, because a stripper had been invited on board. She was a great hit, evidenced by the decided lean of the ship when all the patrons raced to the same side to see her.

We frequently received visitors to the branch, usually to receive a briefing or to brief us on developments. But not all the visitors were there on business. One day we were visited by an old soldier regarded as the 'father' of technical intelligence. Our boss, Colonel Charles, decided that I should be his carer for the day, which meant picking him up at his club, escorting him all day during the talks, presentations and visits that he was scheduled to have, and delivering him back in one piece. Once again, why an Aussie should have the honour of this assignment I didn't know, however the boss had a habit of singling me out for special little jobs. The old gentleman I was to escort was none other than R.V. Jones, Commander of the Order of the British Empire, Companion of the Order of the Bath, Companion of Honour and a Fellow of the Royal Society. Quite a mouthful, and a lot for a simple colonial officer like me to absorb. I frantically tried to read as much as possible of his well-known book, 'Most Secret War: British Scientific Intelligence 1939-1945', which the boss had shoved into my hand. I wanted to sound as if I was smart: that I knew a bit about him, that I knew a bit about Technical Intelligence, or as it was known when he invented it, Scientific Intelligence.

However, I need not have worried. He didn't want to talk about himself, he was far more interested in me, Australia, and the job that we were doing in this modern age of technical 'spying'. It was a pity because I really wanted to learn from this man who had seen so much. His efforts at studying the V2 rocket, the infamous German rocket that rained hell on London, were legendary: including his planning of the British Commando raids on Peenemunde; the test establishment where the V2 was being developed, located on the Baltic Sea on the north-east coast of Germany. It turned out to be a fascinating day and at the end of the talks, the social drinks and the 'special' chats, I was invited, for the first and only time in my life, into a real Englishman's club, *his* club. There, in the shuttered half-light of the sombre, high-ceilinged chamber, where the furniture had probably not been changed or indeed moved since WWII, he proceeded to order double gins from a waiter as ancient as the club itself, and the two of us chatted away as if we were long time spy comrades, who had known each other for years. A memorable day, one that can only happen in a country like England.

Another thing that could only happen in England was to have lunch in King Henry VIII's dining room in the Tower of London. One of my DI60 fellow analysts was a member of a regiment that tracked its history back to Henry VIII's time and had an association with the Tower that allowed its regimental members to use Henry's dining room. Needless to say, Henry's dining room was deep in the main tower and not open to the public, ever! Not only did we have lunch there, amongst his personal battle repertoire including a variety of armoured suits, swords, feathered helmets and other adornments whilst occupying the very chairs that Henry's guests would have sat in; but we were also given a personal tour of the Tower by the Chief Yeoman Warder of Her Majesty the Queens Royal Castle, or Chief 'Beefeater' as he was known to the tourists. We were shown the Beefeater quarters and he told us of the unusual life of pomp and ceremony they lead. Then we were given the inside position to watch the Ceremony of the Keys, which is played out every night when the Tower is closed. Afterwards, we stayed on and had more than a few drinks with the Beefeaters in their mess. It was an honour.

On the theme of Henry VIII, in the basement of the main MOD building, down four floors under street level, there sits a large wine cellar that once belonged to King Henry. It is unusual, to say the least, to take a lift down four floors in the large central London building that was the MOD, then walk several long corridors past many storage rooms and maintenance offices, to break out quite suddenly into a large, open, underground space which is two or three stories high containing the ruins of what actually looks like a small castle. To understand the dimensions of this wine cellar, one has to imagine a large school gymnasium filled with stone parapets surrounding roofless walls of ancient stone rooms. It was a remarkable sight that never failed to surprise me, no matter how many times I saw it. This is where King Henry kept his wines, and drank some of them no doubt, now entombed by the new Ministry of Defence buildings.

The area was capable of holding well over one hundred guests. I found this out because it was the custom of the Australian Officer in DI60, to hold the annual Australia Day function there, and the guest list always went to well over one hundred in number. Dragging cases of Australian wines and boxes of beer down several floors under the MOD was a labour of love, as was preparing the many tubs of ice and the glasses, the vegemite sandwiches and the other goodies. Clearly, Henry did not have such minor issues to distract him from drinking there. In his day, he had not even heard of Australia yet, but I am confident that he would have liked the way Australians entertained in his wine cellar.

Australia Day was not the only time I was expected to play the part of the Aussie expatriate. Australia House was known to have a fabulous free bar, which stocked only the best Australian imported beer. The real deal as we called it. This was not 'Aussie' beer brewed under licence in England, but good stuff all the way from home, free for every visitor to consume. Every month, we Aussies who were posted to England in the various exchange postings would meet and invite our English friends, some of whom complained loudly if they were not regularly invited. It was always a challenge to get home at a reasonable time and in a reasonable state after a night at Australia House. Even the Canadian and US exchange officers would line up to come to Australia House. Their own embassies had no such equivalent. The bar was on the top floor, about ten floors above the Strand, and had large man-sized opening windows. From time to time a few empty cans would 'accidentally' sail over the balcony down onto the street below. Luckily, they were light aluminium.

26 BERCHTESGADEN AND ALWTIC

I n DI60 there were two major intelligence events each year, which defined the branch and the officers within it. They were events where reputations were made or destroyed. One was the Annual Land Warfare Technical Intelligence Conference (ALWTIC), and the other was the British Army of the Rhine (BAOR) and US Army, Berchtesgaden Technical Intelligence Conference.

ALWTIC was managed by the branch. It was a two-week event in London, attended by every TECHINT organisation from the AUSCANUKUS world, for a never-ending series of presentations, meetings, discussions and exchanges. Anyone who was anyone in the world of technical intelligence would be there, even past Australian exchange officers who had previously served in DI60, would be there; but this time representing Australia. It was, for this reason that I attended five ALWTICs: two as a UK Analyst and three as an Australian Head of Delegation. My first ALWTIC in May of 1985, in the first year of my UK posting, was the most nerve-wracking one, but I found that I could hold my own with all the other analysts. As is often the case with these conferences, much of the work is done out of session. It was not unusual to spend every night out to dinner in the West End, or at formal drinks at some embassy, or informal drinks at some club, with a variety of international people. Eating, drinking and talking late into the night for a fortnight was fun, but unbelievably tiring. The Russian spies which were known to frequent the bars round the MOD must have had a field day!

One of our favourite eating haunts was the Chang Mai Mai, a Chinese restaurant of great repute located in the West End of London. Down we would all march, a great blob of suited analysts from all parts of Australia, the US, the UK and Canada, and barge in only to be told 'no loom, no loom' by the proprietor, with a promise that he would come and get us from the White Bear, a pub just around the corner. More beer, followed by a great meal, then off to one of the Servicemen's clubs for even more beer.

On one such occasion, when I was there as the Australian representative, I had consumed far too much alcohol and far too little sleep and had completely forgotten that my presentation was the very first one of the next day. This early in the morning, with alcohol instead of blood racing around in my veins, just speaking was difficult. Imparting technical intelligence to an audience of intelligence analysts was impossible. My brain was in such a state that I could almost feel it catch up to me and bump me in the back of my head every time I stopped moving. I clearly remember one of the American analysts, who had become a good friend of mine, reminding me that the thing I was pointing to on the screen was an artillery self-propelled gun *turret*! The word would simply not come out of my mouth until he whispered it to me! Turret... turret! I can hear him still. As could half the room I think. I scraped my way through the presentation, but my friend never let me forget it. Years later when I left the Army, in the days before email, he sent me a 'signal' as we called it in the military, commonly referred to as a Telex, from the US, as a farewell and a thank you for a job well done, and a cryptic note at the end:

TELEX:
UNCLASSIFIED
ROUTINE
FROM: FSTC CHARLOTTESVILLE VIRGINIA
TO: DIO PASS TO HEAD WEAPONS SYSTEMS LANS
SUBJECT: ANALYST   TO   ANALYST
COMMUNICATIONS

1. (U) ON THE OCCAISION OF YOUR RETIREMENT,
    I WOULD LIKE TO EXERCISE THE MEAGER
    AUTHORITY I STILL HAVE AS THE COMMITTEE
    CHAIRMAN TO THANK YOU FOR YOUR HELP,
    FRIENDSHIP AND COUNSEL (NOTICE THAT
    WISDOM IS CONSPICUOUS BY ITS ABSENCE)
    STOP

2. (U) I WISH YOU THE BEST OUT ON THE FARM
    STOP BY THE WAY THE WORD WAS TURRET
    BEN, TURRET STOP

REGARDS, GEORGE
UNCLASSIFIED

The other big event was Berchtesgaden. This was a conference where analysts were *invited* to attend. The organisers, who were a combination of the British and US Intelligence Branches of the Armies that were stationed in Europe, would only invite those who were regarded as the best in their field. Usually, for us at DI60, this meant analysts who were in the job in at least their second year. The conference was to be held in December and, to my surprise, I was invited to present in my first year as artillery analyst. Lo's parents Cliff and Gwenn were with us, enjoying a round the world holiday, so I decided to be bold and ask the organisers if I could take my family. I didn't accurately define what I meant by my family when I first requested it, and when the organisers approved, I explained that family was not just my immediate family, consisting of my wife and my three children, but my wife's parents, as well. Fortunately, they didn't change their minds.

Berchtesgaden is in a beautiful location in the Bavarian Alps. The conference was held in a large hotel, the Berchtesgadener Hof Hotel, that was now used by American forces as a rest centre, but had once been Hitler's personal hotel, for his guests and for the senior Nazi officers who came to visit him when he was resting at his mountain retreat at the Eagle's Nest nearby. It was in this hotel that he entertained UK Prime Minister Neville Chamberlain. Hitler and Chamberlain signed a non-aggression treaty here in this very hotel, the same treaty Chamberlain famously held aloft in his briefcase as he stepped off the plane on his return to London, to signify that he had achieved peace with the Germans. Chamberlain was known for his policy of appeasement, but he resigned in 1940 when it was clear Hitler would not be appeased. Churchill then became Prime Minister.

My family was very excited! This would be a special occasion for us all. Lo, her parents and the children would be able to do lots of sightseeing, whilst I attended the conference. For an amazingly low charge of $20 per room per night, we were set! They apologetically told me that the family would have to pay $4.50 per meal. Delegates received conference tokens, but they could not provide non-delegates with free meal tokens. Well, we were happy with that. As it turned out, many of the delegates gave us their spare tokens, so the family ate free most of the time anyway. The planning for Berchtesgaden was about to begin. My presentations were the easy thing. The family logistic plan was the challenge.

We decided on a convoluted, complex and, as it turned out, risky plan. Prior to going to Berchtesgaden, Lo and I wanted to go skiing in Flaine, a ski resort in the French Alps, but we thought that Robbie would be too young to enjoy that level of skiing. This was to be the very first time any of the children tried skiing! So, we hatched a plan whereby he would stay with his Nanna and Grandee back at Putney and do some 'special' things. Lo and I and the other two kids would drive to Flaine, with some of our English friends from Coldstream Gardens, have a week's skiing holiday, and then drive to Berchtesgaden. Cliff and Gwenn would fly from London to Munich and catch a bus from there to Berchtesgaden, where we would all catch up together. The only factor that we did not take into account was that Berchtesgaden, being in the Bavarian Alps, near the border with Austria, is subject to snow storms and, from time to time, is cut off by deep snows.

We had to drive almost the length of the Alps to get from Flaine to Berchtesgaden on mountain roads all the way, and when we reached the halfway point, it started to snow, consistently. The snow became heavier and heavier. As darkness fell we started to panic, having driven nearly all day in near white-out conditions. The little Peugeot 305 station wagon was doing so well, but we were up against it! Pushing the little car through hundreds of kilometres of snow-laden alpine roads was a big ask. As the night started to settle, I stopped to put the chains on. The silence of the snow-laden fir forests around the German-Austrian border weighed on us like a blanket. They were magnificent trees. Tall and stately with branches hanging low, snow dripping off and sometimes falling in large clumps on the road ahead of us. Never before had any of us seen so many fir trees, so symmetric, so ghostly!

When I stopped the engine and got out, I could feel the hairs on the back of my neck standing up. Any moment we expected to hear wolves. Involuntary glances over my shoulder betrayed my nervousness. Inside the car there was silence. The whole family understood the potential outcome; stranded in the snow in the Austrian Alps, this crazy family from Australia, who should have known better! Soon it was pitch black. The headlights cut a clear arc ahead that highlighted the inky black all around us. Glistening snow fell on the road ahead, and heavy banks of it could just be made out rising each side of us. Lo remained very quiet, as did Matt and Rae. The two young ones could probably sense our feelings of unease.

We were struggling to continue through the snow when, to our amazement, there in front of us appeared yellow flashing lights. We drew near to discover, of all things, a snowplough! It was slowly but steadily clearing the road ahead of us. It was snowing so heavily that only a kilometre or two behind the machine there was little evidence that it had been there at all. But, tucking in right behind it, we gave ourselves half a chance! We followed, slowly, for a couple of hours, as the machine relentlessly pushed its way towards our target. It was abundantly clear to me that we would not have made it without the snowplough.

To our relief we recognised some road signs and peeled away from the machine, giving the man in the cabin a hearty wave and a big beep! We pulled up at a walled compound, which had huge gates with giant, brass eagles pressed on each gate. Was this the place? The eagles, which were at least two metres high, looked suspiciously like the German WWII Wehrmacht Eagles without the swastika. For a moment, I thought we had arrived at the wrong place, transported back in history somehow. In retrospect, it was not so unusual to see the eagles, this was after all a German Army 'safe place' during WWII. Whilst we stared at the eagles, a small door opened. For a moment I expected a German trooper to step out, but instead a US Marine came to the car, wondering what we were doing there, on this snow-bound road at three in the morning. No Sir, this was not the conference centre, but just down the road we would find it, Sir! They are so polite these Marines. What must he have been thinking? What on earth is that Aussie officer doing carting his family around the Alps at this time of the night, in a snow storm?

So, our trusty little Peugeot had made it. Phew! We soon managed to find Lo's folks and Robbie, who were already there and who were incredibly happy to see us, even in these wee hours of the morning. In his usual bravado, Cliff had introduced himself to the American hosts on arrival and talked his way into the establishment, even without me being there, and he was able to show us our rooms. It was a grand place, that had once been the opulent centre of Hitler's 'getaway', where he relaxed with his mistress and a few of his closest henchman. It gave me goose bumps, again! Cliff and Gwenn had been told that Berchtesgaden was snowed in, the roads closed. No-one was expected to get through for up to a week! The next morning, at breakfast, my boss was even happier to see me, having also written us off because of the snow. But no, not us Aussies! Another story to tell about crazy Australians who won't give up!

The accommodation wing of the Berchtesgaden Hof Hotel, shortly after the family's arrival in conditions of deep snow. This is the hotel used by Hitler to entertain Neville Chamberlain, the British Prime Minister, just prior to WW2.

Photo Lo Lans

The conference went really well. I did the 'work' thing, gave briefings and answered questions and the family went sightseeing, as far as the snowbound roads would let them. This had been Hitler's retreat, his secret holiday residence, the place he courted Eva Broun. There were lots of places for them to see: the Eagle's Retreat itself, the salt mines of Salzburg, the nearby Koningsee, the ski fields. The Americans had constructed a ski resort using the very buildings where, during the war, Joseph Mengele carried out experiments with animals that the Nazis would later try on humans. The sheds that were used to conduct the gruesome experiments now contained hire skis! The resort had lights for night skiing, so this is where we taught Robbie to ski, because he had missed out on Flaine. We quickly realised that we should never have doubted Robbie's ability to learn a physical skill. He took to the snow like a duck to water, at first actually resembling a little duckling as he slid confidently down the slope from the first moment.

There was one other drama that had been averted. None of us had given any thought to the fact that Robbie was on Lo's passport, not on Cliff or Gwenn's, and when they tried to join us and go through immigration at Heathrow, the obvious question was asked: where are the parents? That was such a genuine shock that Cliff and Gwenn must have sounded completely honest, and Robbie would have added his bit. Luckily, they were allowed through.

The week of the conference went too fast. I am confident we set some sort of a new 'car-stacking' record, piling four adults and three children, and all their baggage, into that little Peugeot. By now the roads were clear of snow, or clear enough anyway, and off we went on our way to Munich, from where Cliff and Gwenn would fly to London and we would drive. It had been a great family adventure. The next year I was invited once again to present at the conference. It was however, just before we were due to go home, so Lo and the kids stayed home. That next year's conference was dull by comparison.

27 Adventure Training in the MOD

There was one thing I thought I would be giving up as a result of this posting to London—adventure training. After all, I had been posted to the headquarters of the British defence organisation: the Ministry of Defence. In accepting the posting, I had given up the opportunity to what was potentially the pinnacle of adventure training: an overseas expedition with the Australian Army.

During my time at the 1st Divisional Headquarters in Brisbane, I decided that the best way to spread the word and facilitate adventure training for soldiers across the Army was to create an association any soldier could join, irrespective of rank, unit or location. Its remit: to organise trips at all levels of risk and expertise, to all parts of Australia and overseas. With the assistance of Mal Booth, my friend from 108 and Franklin River veteran, I planned to set up the Australian Army Whitewater Association (AAWA), with a structure based on the already established Australian Army Alpine Association (AAA). The AAA was frequently accused of being an 'officers club', where officers could satisfy their personal climbing ambitions. Looking at the membership of their expeditions, this was not an unreasonable accusation. Unlike the AAA, which consisted overwhelmingly of officers who were graduates of the Royal Military College (RMC), Duntroon,

I wanted the AAWA to be an organisation consisting primarily of soldiers and NCOs. We called our first meeting at Enoggera Barracks in Brisbane during the winter of 1983. It wasn't a conspicuous event. Only seven servicemen came, including Mal and myself. But it was a start! That initial incarnation of the AAWA lasted for just two years and then folded, partly because Mal left the Army and because I was posted overseas, leaving no-one to drive it. Fortunately, some keen adventurers in Army Headquarters re-established it in the early nineties and the new AAWA attracts a large membership today.

Back in the 80s, in attempting to raise the profile of the AAWA, I decided to organise an overseas river expedition. A river in Nepal, called the Sun Kosi drew my interest. Its source is high in the Himalayas. Its snow-fed waters looked ideal for big river rafts. The AAA, which was also planning to do a trip to Nepal, would be a soul-mate organisation with which we could swap ideas and resources. The AAA planned to climb Mount Everest in 1988, the bi-centenary of Australia. I had put a skeleton plan to Army Headquarters, asking for provisional approval to commence planning, and I had received the go ahead. That was encouraging, and my mind started racing with the adrenalin of planning. For me the planning was nearly as much fun as the act itself. But then came my posting to London!

Not for a moment did I consider foregoing such an opportunity, both for my career and from a personal/family point of view. The Sun Kosi would have to wait. As is the way of these grand plans, if the moment is not seized it may slip away and so it was with the Sun Kosi trip. Sadly, the AAWA collapsed while I was away and the provisional approval for the Sun Kosi evaporated.

At first, I thought that adventure training would have to wait whilst I was in the UK. What I had failed to remember however, was the attitude of the British Army to adventure training. In my own work for my Staff College paper, I had declared the Brits to be world leaders in this type of training, but what I had not expected was the extent to which the acceptance and implementation of adventure training had infiltrated the British Army ranks, high and low. For the Poms, it is an accepted part of the training regime.

One day, whilst sitting in my office in the MOD, the door burst open and none other than the boss, Colonel Charles, charged in with an excited look on his face.

'Ben, we are going to do some adventure training. I am taking half the branch this year and the other half next year. We're going to sail across the channel first up, and spend a week sailing the Brittany coast of France. You're an Australian adventurer, I reckon you should come!' he said.

For a moment I just stared at him. Are you kidding?

'Yes, Colonel, I'm in!'

The words all came out at once. I was stunned. I had never heard of such a thing in Australia. Where on earth would half of a headquarters branch pull down the shutters and go sailing? Only in the UK!

It turned out that Colonel Charles was a certified open-ocean skipper, making him eligible to skipper one of the dozens of British Army yachts that were moored all around the coast of the British Isles. These Poms were okay, I thought. Fancy that, having yachts at the ready, all managed by the Army's adventure training organisation.

It was June 1985. Exercise 'Nautical Pongo' (the boss's idea of a catchy name) had just five volunteers. Not even half the branch, but then the yacht only had space for six anyway. It was 33 feet, a sturdy piece of work that looked strong and reliable. The plan was to sail from Poole, near Bournemouth, on the south coast of England, due south across the English Channel and hit the coast of France at, or somewhere near, a French river town called Treguier, which hosts a large quantity of yachts due to its calm water approaches and very yacht-friendly marina. The nearest large port to Treguier is Parros-Guirec, some 15-20 km to the west, but going to a large port was not our plan as a small yacht amongst large ships is not desirable. The crossing was to be a day-night venture. Apart from the Colonel and myself, there were three British Majors: Paul Adams, an Army Engineer with ocean yachting experience who was the first mate; John Goodridge, a Coldstream Guards officer who looked like he never got his hands dirty let alone salty and wet, and Simon Kidner, a down to earth Signals Corps officer who I later discovered liked to have a good whiskey in his spare time. Before we left the dockside at Poole, the four of them got together and doctored my Aussie slouch hat by hanging corks off the rim. They thought it was just hilarious! Being the token Aussie joker was part of my billet in the branch and they reckoned that this was just the way an Aussie should look. They also laughed out loud at my 'pink-zinked' nose: a common thing to do in the strong sunshine of Australia, but completely foreign to these Poms.

We were given the drill by the boss and Paul, and divided into two four-hour watches, the boss keeping himself out of the watch system so that he could be awake most of the time, navigating the busiest shipping lane in the world. We were briefed on how to recognise whether a ship was coming or going at night, by identifying the colour of the lights on the bow and the stern, and on port and starboard sides. We had our own lights, but a tiny sailing boat would barely be noticed by the behemoths that plied the Channel. They were mostly on auto pilot anyway, we were told. Trailing from the back of the boat on a short line was a device that was activated by rushing water to give the skipper the distance travelled. With this information, plus compass headings and the stars, the skipper would navigate us across the Channel. The weather forecast was a little scary, I thought: Force 8 winds were expected. Wasn't that a bit risky? Apparently not, according to the skipper and the first mate, Paul. The winds were 'aft' and so the waves would be following. Okay, I thought. I hope they know what they're doing. We were assigned bunks. Simon and I were given the two bunks up front, right in the nose of the yacht, where the two of us rubbed feet whenever we slept, with our legs cramped into the pointy end of the boat. I thought about the rough seas ahead and wondered aloud at the chance of actually sleeping in those conditions in such a cramped area.

'Don't worry', said the skipper. 'You will be so tired, you'll be out like a light.'

Right, let's get on with it, I thought. We set off in leaden skies and grey water. The south coast of England quickly faded. This was something I had never done—ocean sailing. It was not long before the night came, the ocean darkened, and the Force 8 winds materialised. I was outwardly confident but inwardly less so. My blind confidence in the skipper was beginning to waver at the sight of waves that were as high as our mast, rearing up from behind, barely visible in the dark night until they were almost upon us. These monsters would catch us, slide underneath and roll on, causing the yacht to surf for a moment, and seemingly climb up each wave backwards. The sailing boat would sit for a while, giving us time to look up at the next looming mountain of water that was about to catch us, then slide off the back of the wave, all the while tugging at the rudder to be released and breach. From time to time, Colonel Charles would put his head up from his navigating desk below, look about, then hop under again. Every third or fourth time he did this, he would lean over the edge and empty his stomach, before disappearing to his charts again. I began to wonder how much he had in his stomach; perhaps he was just dry reaching. His face reminded me of one of those cartoons where they would draw a wriggly line for the mouth, to represent a sick person. His mouth was definitely a wriggly line. I have never been vulnerable to sea sickness, and I was glad of that when it was my turn to rest. Making my way from the cockpit to my bunk was no easy task in a small yacht being smashed about, but I made it okay and thought about whether or not I would sleep. Well, I didn't think for long, because in no time I was being shaken by the shoulder:

'Your watch, mates', came John's voice, and Simon and I made our way up top again.

After a couple of shifts the winds abated slightly but the waves were still quite high. It was now utterly black, and frequently heard the thump-thump of large ships: tankers and container ships heading in both directions. I felt tiny and insignificant in this unfamiliar place!

The morning came, grey and sombre. Rain had been falling steadily, but the waves were now much smaller and the storm was abating. The sails had been set to give the helmsman control in the strong seas, but it was time for a sail change. At that moment Paul, who was skippering at the time, spotted a large Russian tanker bearing down on us.

'Okay fellas, now is not a good time to stuff up. He will have seen us. and he will have set his course to avoid us and the last thing we want to do is upset his calculations. A ship like that doesn't change course easily. We probably shouldn't be changing our sail settings, but we need to.'

Simon and I went up to the bow and, nervous about doing something wrong, managed to stuff up the sail change completely! Sure enough, neither of us having any sailing experience, we fouled some line or other. Paul said a few choice words and before long, with the added incentive of the oncoming ship, we had fixed the problem. However, we had lost speed. The long moaning sound of a ship's horn is a dreadful sound when you know it is directed at you. When you are but a tiny blimp in the ocean and the thing that's making the noise is a behemoth, the size of nothing you have ever seen before, bearing right down on you... you feel quite insignificant! Simon and I stood and stared. Paul kept the wheel, his nerve steady. The ship missed us by no more than 50 metres. That is not a big space to have between you and a mammoth tanker which is travelling at pretty much full speed. As the ship charged past, the Captain (or whoever was the officer in charge of the ship at that moment) hung over the edge of the bridge deck (which looked to be at skyscraper height above the ocean) and waved his arms furiously, whilst mouthing unheard Russian obscenities. We waved back meekly.

'Sorry, Paul' was our subdued apology.

'You blithering fools. You'd better learn from this!' he said, grimacing. We had a way to go before becoming better crew members.

As we approached the French coast, the skipper became more and more agitated. It was foggy and hard to spot a coastline or any reference point. Suddenly, there it was.

'Land-ho!' came the cry from Simon.

'Go about... now!' The skipper came back immediately.

We jumped to it. We were soon heading away from the coast, back into the Channel. We were dumbfounded. What was that about? A few minutes later, when the skipper thought we were far enough away from the coast, he ordered a change in bearing and we headed parallel to the coast to find the river entrance that would lead us to Treguier. The tide was dropping very fast and as we settled into the new bearing. It was explained to us that the tides in these parts have the greatest variation of any tides in the world, some 11 to 13 metres. That makes navigation critical, especially around this part of the Normandy coast, where there are huge underwater rock formations that stick up like monolithic spires. Their raggedly sharp, rocky tops are just below the surface at high tide, ready to trap the unwary. No wonder the boss was so wary. He had done a magnificent job navigating us across the English Channel and had missed the target by just a few kilometres, after a stormy and wild night at sea. But we could easily have smashed our keel against one of those rocky outcrops. As the tide began to recede, the tops of the outcrops were revealed. Now we understood the urgency.

Treguier is a delightful little French port. We docked and tied off and the skipper said to wait: the French authorities cannot be hurried. After a while a local French Gendarme wandered down to us and doffed his cap and he and the skipper had, what appeared to be, a very pleasant conversation in French. The Gendarme joined us for a coffee and, after an attempted chat with us all, including me, the obligatory Australian artefact, he wished us good morning and went on his way. This was to be the drill at all the little French ports. What a lifestyle I thought! Is this how the rich and famous live? I want some of that life! We wandered into the town where we found a little bar and had some cognacs and then some lunch. I could get used to this.

One of the effects of the huge tidal extremes was the amazing speed at which the water would flow, either up or downstream, filling or emptying the marina incredibly fast. Those wanting to leave would have to choose their times. As we were having breakfast the next morning, having planned to leave on the changing low tide, we saw a large French yacht, much fancier than ours, impatiently trying to leave whilst the tide was still surging out. The yacht's captain decided to gun the motor and run directly at right angles across the direction of the outgoing tide, past the marina pontoons and into the open channel. A bad miscalculation! Before our eyes, the yacht was swept sideways into the end of one pontoon, then another, then another. Each time the yacht hit a pontoon, it bounced off and the yacht's helmsman would attempt to miss the next pontoon by gunning the motor, failing each time until, at the last hurdle, the yacht was swept sideways into the mooring area, not even making it around the end of the last pontoon. She moored, crippled, with three large gaping holes in her side, just above the water line. Clearly there were some risks associated with this business of being rich and famous and yachting your way along the French coast.

After a few more delightful meals, we headed east to go to Paimpol. This was a somewhat more commercial port which had lots of fishing boats. The Gendarmes were not as helpful here and we had to report to the port authority to have our passports inspected. This was not a place to stay the night, as the wharf had 12 metre high concrete walls that were immersed at high tide, but completely exposed at low tide, showing a muddy and messy looking harbour bottom. Any craft tied up would be left high and dry until the next high tide. We set our course for St Peters Port on the Isle of Guernsey, one of the Channel Isles. This was a fascinating place. Occupied by Germany during WWII, it is a fiercely independent part of the UK with great restaurants and bars that serve 'best bitter' and great meals to all visiting yachtsman. We needed no second invitation. After a good night's sleep, the sail back to the old Blighty was uneventful. The weather was not as rough, but bad enough, with a wind from the north-west which meant we had to beat our way across the Channel. This took time, but it was an adventure to remember.

I must've learned something on that trip because Paul, who'd been selected as a watch supervisor on the British Army's entry into that year's Fastnet race, a large 100 foot yacht, invited me to be a stand-by crew member. A sobering thought. The Fastnet is one of the world's most challenging races, sailing deep into Atlantic Ocean weather. I waited, albeit a bit apprehensively, with my bags packed, until the night before, but the call didn't come. As it turned out, the boat had a crew of 20 but she sailed one crew member short. One of the crew got smashed at the pub the night before and could not be found by sailing time next morning. By that time, it was too late to call the back-up crew, so I missed being part of Fastnet by a hair's breadth. Lo was relieved I missed out, such was the reputation of the dangers of the race, but she understood my disappointment.

Twelve months later it was time to do it again. This time not to France, but to Scotland. The plan was to pick up a yacht at Oban and sail around the Inner Hebrides, taking in Mull and Islay, stopping at some of the many little islands with their quaint, croft-cottage villages. Colonel Charles couldn't go on this one so Paul was the skipper. The crew was much the same, except that a Major Grahame Bone replaced John Goodridge. We caught a train from London to Glasgow to meet up with the Glasgow to Oban train, reputedly one of the world's most scenic train rides.

Sitting in a huge old railway station at Glasgow having, of all things, a cup of tea and scones, I decided to go and find the toilets. With my mind in neutral, I followed the signs, or so I thought, to the gents' toilet, barged in confidently and started reaching for my fly zipper... except it wasn't the gents. I was in the ladies' toilet, and it was packed! Bemused faces turned and stared at me! Damn, now I was flustered. I mumbled a quick apology and turned on my heels to make an escape when... crash! Simon ran headlong into me! He'd followed me blindly, trusting that 'Lans knew where to go'! Wrong! Now there were two of us in the ladies'. More mirth and laughter. Scrambling to get out we fought for the door and then, Paul burst in, nearly knocking all three of us into a heap on the floor! Without thinking he had followed Simon! This was complete chaos! Some women stood dumbfounded, others were doubled over with laughter, watching first one, then two and now three men panic and scramble unsuccessfully for the door! A roar of laughter was all we heard as we hurried out red-faced. Paul and Simon didn't let me live this one down too quickly.

'Don't you bloody Australians know how to read a sign?'

'Don't they have segregated toilets in Australia yet?'

The train journey to Oban was beautiful, as promised. Misty and mysterious mountains surrounded the line, with the train stopping at tiny magical highland stations where Scots alighted and headed home. I wondered what it would be like to live here in these legendary highlands. Perhaps their ancestors had been highland clan warriors.

Before sailing from Oban we decided to partake in a tour of the famous Oban Whiskey Distillery. Despite the hour being only ten in the morning, the tour offered free whiskey shots. We didn't need prompting. This was the awakening of my taste for good whiskey, and it made for a warm start to a cold day's sailing. After one last stop to pick up some smoked herring, we were off.

Our first stop was Puilldorrain, then we sailed across to the Isle of Mull and into a little loch called Loch Spelve. This was the first of many tiny settlements that we visited around the coast and islands of Scotland, often comprising a few little crofter's cottages, occupied by a handful of residents. We dropped anchor and rowed ashore, to be met by most of the village I'm sure. They spoke a foreign language... Gaelic, as foreign to us all, even to the British officers, as was Russian or Mandarin! But we didn't have to translate the spoken word as the body language said it all. Come and have our whiskey! This was lesson two in my whiskey drinking education. I was learning fast! Every village seemed to distil its own whiskey, which they would serve in their own little pub. There was always a pub. If there was more than one house in a village, one of the two houses would be a pub! Oh, and don't ask to mix anything with it: no water, no ice! We may not have spoken their language, but we understood quickly enough the sanctity of their whiskey.

The sailing experience here was much different to sailing the open spaces of the English Channel. As we beat up the Sound of Mull, the cold wind ripped around our ears and noses. Sailors from up here must be tough men to put up with these conditions every day. That night, we sheltered at Tobermory on the Island of Mull, one of the prettiest harbours I have ever seen. Then Oronsay on Loch Sunart, for some more whiskey, and on to Gometra on Ulva, a small island where we planned to go ashore. The wind had come up and was blowing across the island, over the town and out of the loch, pushing our boat out into the ocean. Paul decided to stay on board as it was rough, and he worried that someone might be needed on board to look after things. He briefed us in great detail.

'When you row back, the wind will be behind you. Don't come too fast as you may miss the boat and be swept out into the ocean and you'll be on your way into the greater Atlantic! Give me plenty of warning by yelling and I will throw you a line.'

That seemed a little extreme, as we were not really facing the open Atlantic!

'Aye aye, cap'n,' we joked, and set off for another whiskey-drinking adventure.

We drank. Did we ever! Picture three rather overly-jolly officers, paddling a tiny, tiny little rowboat, with its gunnels only just above the water line and waves uncontrollably splashing in, all yelling at the tops of their voices:

'Paul, Paul... here we come!'

We came all right! Straight past the boat! By the time Paul was alerted, he had little time to throw a line, which we would have been too drunk to catch anyway, and off we sped in our sinking rub-a-dub-dub tub, propelled by a strong, gusty wind, straight into the open ocean to who knows where? In the pitch black of night, a howling wind and water splashing over the side, none of us were sober enough to grasp the extent of our danger. I can't remember exactly what Paul said; none of us could recall, really, but it must have been choice. Hauling the anchor and starting the engine, he chased us before we completely disappeared into the emptiness of a huge cold ocean, in this tiny unseaworthy dinghy.

Sailing from the Isle of Mull heading to Staffa.
Photo Author

The island of Staffa on the horizon.
Photo Author

A highlight of that trip was a visit to the island of Staffa, where Fingal's Cave is located. This is the cave that inspired Mendelssohn to write his Hebrides Overture, a stunning piece of music that copies the deep sound of the waves smashing against the end of the cave as the basis for his overture. The shape of the cave is such that the large entrance catches the waves that roll in from the Atlantic, only to squeeze them as the cave narrows, causing them to smash against the wall at the end with a crescendo of noise that reverberates all the way out again. It is an inspiring sound, as regular as the ocean's pulse.

From there we dropped south to Scalasaig on Coronsay, and then skipped around Jura, through the Sound of Islay before doing a spinnaker run up the Sound of Jura through the infamous Gulf of Corrywreckan to Crinan, infamous because many a fisherman had been caught out here on a treacherous piece of ocean called the horizontal waterfall, where the fast tides race across a drop in the ocean floor setting up whirlpools and currents that may trap the unwary. After seven days in and around the dark and foreboding waters of the Inner Hebrides, we sailed into Oban again. What a magic experience it was! Some of the little crofter's villages that we had seen are as remote on the west coast of Scotland as any of our remote settlements in outback Australia. These people speak a language that is dying fast, or at the very least, being diluted. Their way of life is under threat as medical help and education for their children is difficult to obtain, and the cost and logistics of trying to eke out a living in a remote village visited infrequently by mail and cargo boats, is enough to make many of them quit and retreat to the mainland of Scotland. I felt that I had been treated to a snapshot of a special way of life, welcomed again and again by special people who could barely communicate with me, except by the universal sign language of gestures and smiling faces... and whiskey. Not just any whiskey, but some of the best peat whiskey in the world.

28 Living in London

Our time at Coldstream Gardens was marked with events that were everything from hilarious to incredulous, from warm and welcoming to downright insulting, but mostly good fun and always a good experience. It was an unusual place to live because we weren't used to being part of a 'married patch', where dozens of families were jammed into a relatively small area. Most of the officers worked in the Ministry of Defence and most of them were of rank higher than Major. Thus, I was not only at the bottom of the pecking order as a Major, but as a 'colonial Major', I hardly rated at all. Many British officers viewed Australian officers as their professional equal, a view that had been developed over many years of Commonwealth cooperation in war and peace, and latterly through many exchange postings. Conversely, there were some in the British military who, through lack of contact, ignorance or involuntary adherence to an inbred class system, regarded officers of countries like Australia, New Zealand and Canada as slightly quaint, colonial and certainly not 'one of us'. I hasten to add this view was not the norm in most of the regiments of the British Army. However, this was not just any regiment that we happened to be sharing quarters with. These were mostly officers from the Coldstream Guards, a regiment that goes back hundreds of years, that has done more than its share of Buckingham Palace duty, and that feels itself to be a cut above even other British regiments.

Living at Coldstream Gardens was an experience. Every morning the officers would walk past our kitchen window on their way to the tube, with their bowler hats and umbrellas, in the uniform of the upper middle classes. Military personnel were not allowed to wear uniform on public transport in those days. The IRA were very active in England and to be wearing a uniform on the street was to mark yourself as a target.

The house itself was a modest but comfortable four-bedroom, three-storey, single bathroom, semi-detached place. Incredibly, our bathroom was on the top floor and, in true British fashion, had no shower. Being on the third floor, there was little water pressure, certainly not enough for the water to rise just a few feet more to permit a standing shower. So, when we connected a hand-held shower to the bath tap, we could only wet ourselves by lying down in the bath and running the shower head along our bodies horizontally.

The house was not large but so well placed near central London that it was a highly sought-after location. These estate houses were very well designed and situated to avoid windows that overlooked the neighbours' windows. Most of the houses looked into a sizable common area with trees and grass and space to run, and a small children's playground.

East Putney Tube Station was only about 10 minutes' walk away, and the ride from there to Embankment, which was the station nearest the Ministry of Defence where I worked, was only about 25 minutes. Those tube rides were memorable. Air circulation was not a feature of the tube design in those days, and in summer the air was so thick with human odour that the smell actually stuck to my clothes and the inside of my nose for a long time after each ride. Luckily, East Putney was about half-way between the terminus at Wimbledon and the city, so in the morning there was always a chance of finding a seat. No such luck on the way home though, when I would get on in the midst of the London peak-hour crush.

I learned to read *The Times* standing up, the paper folded into eight, with the particular article of choice right in front of my eyes, so that I wouldn't have to look at any other commuter in the face. Eye-to-eye contact on the tube was a big no-no! It was an acquired skill to hold on with one hand, swaying just enough to go with the flow, but not be suspected of intruding into anyone else's space, which on a tube is just a few very important inches. Sometimes, because passengers were pushed together so firmly that nobody could move, we were forced into the funniest, most compromising positions. I remember one hilarious situation where I was holding one of the upright supports when the tube jam became so tight that this poor girl found herself pushed hard against my hand with her bosom. Here we were, our faces inches apart, my hand right there, held tightly to the post by her enveloping breasts (and she was quite well endowed), but there was nothing either of us could do about it without drawing attention to ourselves. So, there we stood, trying not to catch each other's eye. If we had strayed beyond that convention I think we both would have burst out laughing. Perhaps we should have.

The schooling in England was excellent. Courtesy of Australia House and the Australian taxpayer, we were able to send our children to public schools, known in our Australian lingo as private schools. Everywhere in England these appeared to be of the highest quality. Matt and Robbie went to a preparatory school called Willington, which consisted a large four-storey house with multiple rooms, cellar and a very, very small playground. Willington lays claim to Lawrence Oates as one of its old boys. Oates was the member of Captain Robert Scott's failed expedition to the South Pole in 1911, the man who walked off into the snow and sacrificed his life in order to preserve food for the remaining two members of the expedition. It was to no avail as the others died within a few days. But Willington was proud of its traditions. Mind you, we had been warned off that particular school by one of the fellow Coldstream Gardens residents, a lady by the name of 'Snooky'. Snooky considered herself very much upper class and warned us that Willington:

'...did have rather a preponderance of brown faces, not the ideal Anglo-Saxon mix, you know!'

Snooky went on to show her true colours later when she made some very derogatory remarks aimed at one of young Matty's South African friends, so cruel that she made the poor boy cry. Lo later re-named her 'Snooty'. Needless to say, we had little to do with her and the few like her. Most of the residents of the patch were friendly and accommodating people.

We soon became engrossed in the daily routine of life in London. The house at Coldstream Gardens was chaotic every morning. All the usual panic in any family with three young children would occur, amplified by the fact that the house had three stories, thus the shouting was even louder, and generally accompanied by much running up and down stairs. Lo would get the gang ready for school which was not an easy task. The dress rules for the little preparatory schools they went to were diabolical. Matt would race about looking for all the bits of uniform that he needed for the day, sometimes in a panic. The shoes alone were a challenge! Leather tie-ups to use to and from school, sandals for inside the class room, plimsolls for playing outside, wellies if it was raining and all sorts of sport shoes for playing anything from cricket to football to athletics. The poor kid. Whilst they had breakfast, they'd watch the procession of Coldstream Guards Officers filing past our kitchen window and not long after that, Lo would set off with Matt and Robbie to Willington in Putney, and Rae to her school: 'The Study' in Wimbledon.

The custom at prep schools in England was to eat a hot lunch at school, and so that is what our kids did. Steaming hot stews or sausages or rissoles, with thoroughly well-boiled vegetables and potatoes. Good solid—almost stolid—English fare! The boys thrived at Willington; they loved it there. Robbie was so young that he developed a real little 'posh Pommie' accent almost immediately. His daily play space was the size of half a tennis court, ringed with high wire so that the little fellas couldn't escape. Robbie took to the school very well and really blossomed, in spite of the fact that Lo discovered, right at the start of the posting, that he was entirely deaf in one ear. Her suspicions were confirmed when she asked him to answer the phone

one day and he said to her there was no one on the phone, when in fact Lo could hear my voice saying 'hi' to him. The subsequent tests revealed permanent damage to the vital nerves of his left ear. In spite of that drawback, he was a good student, our Rob.

Both boys benefitted from the strict discipline at Willington, which set them up for their later schooling. Matt never had any doubts about the need to complete his homework ever again. I was never sure what punishment Willington metered out for failure to complete homework. I think it was just the displeasure of the master, nothing physical, but he learned to do as he was asked at school, and it stood him in good stead for the rest of his schooling life. Matt had quite a hard time of it at the start because he had to add Latin, French and a much higher level of maths to the curriculum that he had been used to in Australia. The poor fellow was really under the gun. Luckily for him, and for us, his grandfather, Cliff, stayed with us for about five months. Cliff had studied Latin at school. Frequently you would see Matt and his Grandee doing Latin homework together. Matt can be proud of his achievements and set a great example to his sister and brother.

The school was very active in sports and Matt almost immediately became something of a school champion, winning the cross country on his second day at school. He went on to become a good athlete and all-round sportsman. Lo's background as a Physical Education teacher came to the fore and she loved helping to coach Matt and the other children. Her input on the sports field became a regular feature of the boys' two years at Willington. Lo helped train the athletics squad at a London athletics track and a favourite story of Matt's was that he trained for athletics on the same running track and at the same time as Daly Thompson, the champion British decathlete of that time.

Rae went to a girl's school called The Study, where she also did very well, particularly in such things as gymnastics, dance and music. She was chosen to play solo recorder for the annual Christmas service in the school's grand cathedral-like church. The haunting sound of our little Aussie daughter playing 'The Little Drummer Boy' was a moment to remember. She made some nice friends there that she remained in touch with for many years. The Study was an all-girl's school situated in leafy Wimbledon. The school had some fine teachers who came to appreciate our little girl's talents with dance and gymnastics, as well as her academic prowess. When we first enquired for a placement, we were told by the school secretary that there was simply no room, however we might discuss with the Headmistress, a placement in the future. On no account were we to take our daughter along for the interview, as this might unfairly persuade the Head, she told us. We purposefully *did* take Rae along, and the Head was immediately taken with our little Aussie girl, and she was offered an immediate placement. I still remember the secretary's exasperated look!

Raewyn benefitted most from the cultural life in London. At the young age of seven she was old enough to enjoy the theatre, and she just loved every minute of it. London is the home of musicals, pantomimes and children's theatre and I, as a soldier, was able to get very cheap seats for almost all the children's and adult's shows, including such blockbusters as Andrew Lloyd Weber's *Starlight Express*, *Cats*, and Tim Rice's *Chess*, plus lots of pantomimes. This was because of a delightful English tradition that soldiers, fireman, ambulance officers and police members were entitled to reduced cost tickets in the West End. Tickets for 50p, what a steal! *Starlight Express* was the first real big time musical the children saw; they were stunned and immediately hooked from their front row seats. Because the West End shows were very expensive, children were rarely seen with the result that as soon as the actors identified that there were children in the very front row, they played to them! Lo took the kids to many London productions and any family or friends who came over from Australia (and there were lots), were carted off the West End to see a show.

Australia House approached us to allow the three of them to act in advertisements that would screen in Australia. Matty scored a role with Penelope Keith, starring as her little offsider in a Heinz soup commercial.

He was more proud of his first pay packet than his starring role fame! He had to ride alongside her on Wimbledon Common in an apparent 'rain storm', then go inside their home to eat hot soup and speak his only line... 'it's got chunks in it'. The only problem was the soup was stone cold, the steam being artificially produced by dry ice! Anyway, he managed the part, and we decided not to tell anyone in Australia that he was in this ad, just to see their surprise. That turned out to be a bit of a mistake. Since absolutely nobody back home expected Matt to be in an ad on television, his one performance as an international actor went entirely un-noticed.

Like all good Aussies abroad, we travelled as much as possible in Europe and throughout England, Wales and Scotland whenever we could. Renting little crofts and cottages in places like The Forest of Dean, and the highlands of Scotland or the mountains and valleys of Wales. On occasion, we would meet up with Lo's parents, who were doing their own touring, making use of the time they had with us at Coldstream Gardens. In Scotland, our kids were so keen on swimming, even the freezing cold water of the Scottish Highlands was not enough to stop them, to the astonishment of the locals.

On one occasion, we arrived a little late to take up a cottage in the highlands: the sun had set, the weather had closed in and the track was flooded. This was what we might have expected in Australia but not in Scotland! We seemed to be in the middle of nowhere but, after asking for advice at a huge castle-house in the dark, we were given directions. I felt distinctly uncomfortable, banging on a huge set of double doors at a castle entrance where all I could see was a faint light on in a high turret, with the rain pounding away at my back and lightning flashing. It was surreal. I almost expected Frankenstein to answer the door! After a few more loud and long knocks, one large door opened creakily followed by the ubiquitous... 'Yaaaas'. Instead of Frankenstein, a kindly woman's face appeared. She explained the way, the number of gates, how to open them, how to avoid the loch, and off we went into the darkness again. We found our crofter's cottage, not without a few scary moments.

The Scottish Highlands can seem as remote as anywhere in the world! The cottage was gas-lit by individual gas lamps that depended on fragile mantles, one of which managed to break as I was groping around looking for a light switch. But after the trials of the night just to find the place, we were grateful for the shelter and soon made ourselves dry and warm in front of a roaring open fire. The little cottage seemed ideal, until the next morning. It was a bright, clear and sunny highlands day, but when Matt opened the door and stepped outside, we suddenly realised why the highland fling had originated in these parts. We could actually *see* the miniature-sized midges rise from the grasses in front of the door and turn towards the cottage.

'Quick Matt, shut the door!' I yelled, to no avail.

As soon as the midges hit the door, they started to search for a way through, around and underneath, coming inside though every crack they could find. We had to learn fast how to deal with these pests. The best antidote was to simply stand in the strongest wind you could find.

We had some great trips through Europe, particularly France. Our mid-sized Peugeot 305 station wagon was quite adequate, but it let us down in France—just once. The rear suspension started making clunking noises, until eventually the car broke down with an almighty cracking noise. A local farmer, whose daughter spoke English, came to the rescue after being alerted by his daughter, and pulled the car up on his flat-tray truck to take us all into the nearest little coastal town, a place called Treguier. It was the same little coastal town that I would later visit on my MOD sailing trip. There, Lo, who had a smidgeon of high school French, explained to the mechanic about 'la clunk' in the 'derriere' of the car. He thought that was hilarious! He fixed the car but couldn't fix 'la clunk'. The clunking came straight back, but we managed to crawl home to England.

On that same trip, we went to Holland where we visited my birthplace, Haarlem, and Amsterdam. When we crossed the channel into Dover again and were waiting for customs entry, I was questioned by the official as to where we had been.

'Paris and Amsterdam,' I said casually, looking rather scruffy with a three-week beard and a car stuffed with bags and barrels and people. And we were Aussies to boot!

'Why don't you just pull over here, Sir. We might just take a look.'

When I pulled into the side lane indicated, Lo saw the previous car that had been singled out. It was an old Bentleigh, and it was being taken apart door panel by door panel, with the contents being carefully spread over the ground radiating out from the car. We quickly decided that fate was not for us, and Lo reckoned I should use my position at the MOD to get us out of this. In any case, we had a fair bit of extra alcohol in the car, well over the limit. I went up to the customs official and showed him my MOD pass, which showed the level of security clearance that I had, way above what he had himself. He looked at me in disbelief. What was this scraggy Aussie doing with this level of security clearance? But there was no denying it. Phew! Through we went with just a brief and polite lecture on taking too much alcohol, but without the detailed search.

One of the strangest things about Coldstream gardens was that there were almost no children there. It was not because the officers and their wives who lived there had no families, it was because they sent their children 'away'. This was a term we were not familiar with.

'Away? Where?', we asked.

'Away to School!' was the incredulous answer.

Boarding school, of course! From the tender age of eight, most of the children there were sent to a boarding school somewhere, to come home only on special weekends and holidays. It was a long-standing tradition in the British middle and upper classes, though totally foreign to us. As a result, for much of the year, our children had the 'patch' to themselves. In the tradition of good Aussie families, we would play things like footy and cricket out there on the grass, where everybody could see us, with me quite often wearing my sarong. This would result in some strange looks and the odd curtain being drawn; 'Can't have this going on in front of our noses, you know!'

We absolutely loved our time in London. Lo and I made friends and the children made friends and everywhere we went we were welcomed. After all, every English person has a relative in Australia... 'you know, in Sydney, her name is, do you happen to know her?'. My replacement came all too soon. My good friend Major Peter Overstead and his wife Bronwyn arrived and we showed them around London. We took them to the West End and to the Barbican to attend a concert by Yehudi Menuhin, the world-famous concert violinist, and they were astounded at the life we were leading—and their life to come—with all the things you could do here in this wonderful city. We gave up our Coldstream house to them and Australia House moved us into the ultra-luxurious Grosvenor House Hotel in Park Lane, in the city, where we were given a double suite. We stayed there for ten days on allowances so generous that we were able to save thousands of dollars for our upcoming trip around Australia.

Our last fling was a Sunday dinner at a local pub with all our friends, and we were off home to Australia. We weren't ready to give up this good life just yet, and were silent in the taxi as we drove through the city, seeing off a wonderful posting and, in our minds, saying goodbye to friends who would very soon be a world away. The harsh Aussie accent of the Australian hostess on the QANTAS plane at Heathrow, suddenly brought us right down to earth. As we left there were tears of sadness, but also excitement at the prospect of seeing family and our beautiful home country.

Whilst in England and Europe, Lo and I realised that we didn't know enough about our own country. So, we decided to go back and do a trip around Australia for the best part of a year.

I'd informed the Military Secretary, the good Brigadier Ian McGuiness, my ex 4th Field CO, and he agreed to give me a year's leave, being my long service leave at half pay. He did say however, that he had planned to promote me to Lieutenant Colonel on my return, but that he would now not do that, as I was taking this lengthy leave. I could see the logic, but I was disappointed. Nothing was more important than my family though, so I resigned myself to having to wait the extra time. As it turned out, because I went on that leave, the posting cycle was thrown out of whack and I had to wait 18 months for my promotion to Lieutenant Colonel.

No regrets, though. We had a blast!

29 Home to Australia

Coming home to Australia was a reality check for us all. Career wise, the TECHINT posting was one of the best postings I had ever enjoyed. The freedom and responsibility in the MOD was much greater than I could expect in a similar position in Australia. There was an interesting work ethic amongst the British officers, a determination to get things done and done right, whilst on the surface they displayed a 'devil may care' attitude. A 'casual external outlook, belying a caring professional approach' was how my ex CO in Townsville described my work ethic. And it was a work ethic perfectly suited to the MOD environment, one in which the responsibilities and outcomes are clearly defined, but a person is allowed to find their own way of achieving.

The relationships between officers of varying seniority also appears to be more relaxed than in the Australian Army but no less respectful. This is exemplified by the delightful way in which junior officers address senior officers—using their first name preceded by their rank, like 'Major Jim' or Colonel Bill', unlike Australia, where everyone senior is just 'Sir!' It is clear that the British have nothing to prove. They have 'been there, done that' for a long time in the history of the world. Proud traditions, sometimes mixed with an irritating dose of pompousness and an air of superiority, are carried forward with a swagger, in a military culture that displays clear and uncompromising confidence in itself. The Australian military culture, I realised, is different—more down to earth and straight forward, less pretentious and with that mix of larrikinism. Amongst Australian soldiers and junior to middle ranking officers, there is great camaraderie. But take the officers away and put them in a headquarters like Army Headquarters, and attitudes change. Overly close supervision of staff, lack of personal expression, and an over-developed sense of pleasing superiors seems to take over. This style of leadership does not suit me. I need to be able to express myself, apply what I think is the best way ahead within the guidelines as given, and prove the outcome with results, not with endless platitudes and promises.

It was with mixed feelings that I returned to Australia. As one becomes more senior in the services, the relationships with soldiers change, and life as an officer becomes one of planning, developing and writing, creating policy, conducting studies, advising senior officers, staffs and politicians, and conducting briefs. I thought I was not quite ready for that world, unwilling to leave a professional world where I could work closely with soldiers out in the field, not in an office.

For the family, there were also big changes. From attending exclusive, small preparatory schools in London, the children were now thrown into good ol' Aussie semi-country co-ed schools. Knowing that we were going to take a year off to travel Australia, we moved in with Lo's parents at Glenorie, on the outskirts of north-west Sydney. From here we planned our big trip. The idea was to take our Nissan Urvan and deck it out for camping, attach a trailer for the gear, and set off with no particular planned route other than a vague idea that we wanted to see the inland, the north and the west of the continent. In Australia, that is one very large area, considerably bigger than all of Europe. We approached the NSW Education Department for advice about schooling, and once they realised what we were doing, that Lo was a teacher and we appeared to be capable people, they said on the quiet:

'Why don't you just teach your children yourself, as that will be much more efficient than trying to communicate with us by sending lesson plans and assignments back and forth from all over Australia?'

So that's what we did.

For peace of mind, we added the 'luxury' item of a high frequency radio to our kit, which we could use on the Royal Flying Doctor Service radio network. I figured that my experience as a signals officer in Singapore was more than enough to allow me to operate one of these radios. This meant that, in those days before mobile phones, and when phones on the satellite networks were prohibitively expensive, we could still be in touch from anywhere in Australia to anyone in the world. We planned to call up the nearest Flying Doctor station from wherever we were and ask to be connected into the telephone network. This way we could talk, via a radio wave bouncing off the ionosphere, into a telephone line connection to our friends and family. Raewyn talked to her school friends in Glenorie from somewhere around Ayers Rock. That was quite an event in those days.

1987 was a truly memorable year. Our children, fresh from the experience of English schooling now learned to deal with mum and dad around them all day and conducting lessons in the heat and the flies of country Australia. We found it less difficult than anticipated, because once you are in a one-on-one relationship with your child, educational concepts are comparatively easily explained. There's no waiting for the slowest learner in the class. Lo took Robbie, teaching him to read and write, which was challenging as she had no young primary school experience, despite being a teacher. I took the two older ones, and we did maths, geography, Australian history and English. This was not too hard because the levels were still very basic, and I used the treks we went on, the places we saw, and the libraries along the way to link everything together; by getting them to research, write essays and poetry about it. All three of them submitted contributions for publication in *KidZone*, a young people's magazine, and Matt was called their 'roving young reporter'.

Towards the end of that year, it was time to think about the real world again, where we would have to work and live in a house and where our world did not consist of the car, the camp and the road. We were in Western Australia in late November when I realised that my leave was nearly finished. When you take off on a period of leave that is about 11 months long, the actual 'end of leave' date is not something that is prominent in your mind. We raced back across Australia and in no time, I found myself standing in the office of Lieutenant Colonel Al McClelland, who was to be my boss for an interim period of six months whilst I awaited my promotion, and who had been my predecessor at 108 Battery. He didn't recognise me with my 12-month growth!

I'd been posted temporarily to the Directorate of Artillery, awaiting the slot in the Defence Intelligence Organisation (DIO) to be vacated. I was to take over from Lieutenant Colonel John Cox, who had also preceded me in London, and who now preceded me in Canberra. After John, I would be the Head of Army Technical Intelligence in Australia. This was a far smaller organisation than the one in the UK. I wouldn't be taking over until mid-year, so I was basically marking time for six months.

During that period, I was asked by the Director of Artillery, Colonel Tim Ford, to attend a Four Nations Quadripartite Conference in Woolwich, London. Woolwich is a long-standing Artillery Barracks. The Quadripartite Conference was held every year, hosted by Australia, Canada, the UK or the US. That year 1988, was the turn of the UK, so lucky me, I was off to London again! The purpose of the conference, and the Quadripartite agreement as a whole, was to standardise the artillery operating procedures and design standards of the four nations. This would facilitate better inter-operability and international understanding in times of joint operations. Colonel Ford handed me the files, so that I could familiarise and prepare myself.

It was quickly evident that these conferences were the domain of the professional bureaucrat. I looked at the minutes of previous years' meetings...

and they said it all. The same topics for discussion just kept surfacing and re-surfacing. Each time, one country or another would be tasked to investigate and report in 12 months' time, and every time the report would consist of baffling, sometimes indecipherable waffle, followed by endless formally seconded motions, or excuses really, and a series of actions to be taken in preparation for next year's meeting. Always another meeting, always another recommendation, never a solution.

What a waste of taxpayer's money! But hey, why should I be the one to buck this tradition of a 'swan'? In any case, in the back of my mind I could see that I could actually do something useful on this taxpayer's jolly. Thinking of my previous posting in MOD Techint, and my next posting in Australian Techint, I could take the opportunity to attend the ALWTIC conference, which was to be held a few days after the Quadripartite Conference finished! Both were in London. It promised to be an exciting trip, seeing some old UK artillery mates who would be at the Quadripartite conference, and then seeing my Techint mates at the AWLTIC, as well as the personal friends that Lo and I had made during our stay in London. I approached Colonel Ford, and he readily agreed.

I arrived in London after a business class flight on QANTAS, feeling quite refreshed, and settled in at the Woolwich Officer's Mess. I was surprised to find civilian managers, cooks and stewards. The place, with its hundreds and hundreds of years of history and rich tradition, was being run like a hotel. I was still used to seeing a Mess Sergeant running things, with Catering Corps Cooks and Stewards, but the Brits were starting to do away with that system, believing that every man in uniform was supposed to be trained for deployment, not working full time in a mess at home. The test of time was yet to prove how successful this approach might be. The obvious problem was that, if you don't have 'peace-time' employment for your cooks and other Catering Corps staff, how would you keep their skills up and where would you get them from when your defence force is deployed?

The most remarkable aspects of the Quadripartite Conference were the mess drinks and dinners. These were all 'unmissable' and where the true value of the conference lay: the friendships and lasting contacts created between officers of the allied Armies. The conference lasted about a week, and on one of the nights when there was no official function, a rare occasion, I decided to go and visit some of our London friends. This was a time of high security in London, because the IRA was active and dangerous, having already proven to be capable of severe acts of terrorism, such as the terrible slaughter of a column of men and horses of the Royal Regiment of Horse Guards, and the assassination of Lord Mountbatten.

After a big night out, I caught the tube back to Woolwich and walked back to the mess. It was late at night, well after normal pub closing, but as I approached the barracks I could see over the old stone fence, the one that had the sharp rocks on top which was designed to keep intruders out, that there were still lights on in the officers mess. All entry to the barracks was via the main gate, where every visitor was checked, identified and, depending on the level of access granted on the visitors' pass, was either allowed through or delayed until an escort could be fetched. All vehicles were searched and checked underneath with mirrors for bombs under the chassis. I had a pass so no problem. However, in my naivety, and not being used to such strict security measures, I decided to climb the wall, which was only about a man's height, and walk straight to the mess across the parade ground, rather than walk the extra kilometre or so around the block to the front gate. So, in my best suit, and having had a few whiskeys, but not too many, just enough to make me brave, I confidently jumped the wall and strode towards the mess.

I had only take a few steps when:

'Halt!' came a strong, distinct cockney voice.

'Place your hands on your head, stand very still! Next to me stands a soldier with a loaded rifle pointed directly at the middle of your back. I have a large guard dog on a leash. If you move your hands I will order him to shoot and I will release the dog. Do you understand?'

Bloody hell, did I understand? He could have been speaking Russian and I would have understood the intent of what he was saying! I blabbered something, put my hands on my head and waited, what else could I do. Here we go, me in the Military Police lock up for the night, what will the boss say? 'You idiot Lans!' I could hear Tim Ford now. Then the voice again:

'We are approaching now. Stay very still. I will search you for weapons!'

It seemed to take forever but I could sense a dog right next to me, looking straight up into my face and poised to jump if I even looked like moving. A beautiful German Shepherd I realised, but for once I did not feel like patting that dog in the way I normally would.

I felt my arms being dragged around the back and pinned but I was still looking away, scared to even turn my head, having not yet laid eyes on either of the guards. I felt hands searching me and patting all the places a weapon or explosive could be hidden, then my wallet was lifted.

'You are an Australian officer?' came the somewhat bemused voice.

'Yes,' I blurted out, 'I am attending a conference here and was on my way to the mess.'

'I see that,' the voice again, as he spun me around.

There staring at me, from about 12 inches away, was the barrel of a rifle being held by a serious looking soldier, aimed straight at my chest, whilst his eyes were looking directly into mine, ready to detect any flicker of warning, should I even think of making a move. I shifted my gaze slowly to look at who was doing all the talking and I saw a smartly dressed Military Police Corporal, who by this stage had actually softened a little and gestured for me to relax my arms. I thought I could detect a glint of sympathy in his eyes.

'So, you thought you might take a shortcut... Sir?' he asked sarcastically, the 'Sir' coming mockingly after a suitable delay.

'Okay, so let's see if you're really who you say you are. Walk over to the mess window and look in. I want you to identify someone who can vouch for you.'

Uh oh, I thought, please let there be someone I know! Yes! There was one of the officers I knew! Some Canadian whose name I actually remembered in the heat of that moment. In no time, the Canadian was outside wondering what on earth this was all about. By now the mess window was full of curious faces peering into the dark to see what this stupid Australian was up to. Within moments I had been identified, lectured to by the Military Policeman for being so stupid, reminded that England was in a state of war against the IRA and that I was not in Australia now, and summarily dismissed as a would-be intruder. Then off they went: corporal, soldier with rifle and dog, into the darkness to carry on securing the barracks. I was impressed, although I was still shaking, even more than before.

'Come into the mess, Lans, you need a drink!' I could hear someone saying.

Soon I was in the warmth of the bar, the centre of attention, relating my story to all. It had been the most exciting thing to happen at the conference, by a long shot! I felt a bit humiliated by the experience. However, a few beers and a good laugh, and some embellishments from the course officers, soon helped to make this a story worth remembering.

The Quadripartite conference finally ground itself to a standstill and I relocated myself to the United Servicemen's Club, just off the Strand near Trafalgar Square. It was right opposite the Metropole Building of the MOD, where ALWTIC would be held and where I would support Lieutenant Colonel John Cox in his dealings with the AUSCANUKUS community as the Australian representative. The usual long hours of meetings—followed by drinks and dinners—went on for two weeks, on top of the social events with our London friends on the weekends. This led me to a state of exhaustion, in which I was desperate for a good sleep. I could barely stomach the thought of another pint of the best British bitter. But soon enough another ALWTIC had been successfully completed, and, after a stop off at Hong Kong on the way home to top up on goodies for Lo and the kids, I was home.

The rest of my posting at the Directorate of Artillery was uneventful. I was just waiting for John Cox to move on from his appointment so that I could take over. At last, in the middle of the year in 1988, it was my turn and, newly promoted to Lieutenant Colonel, I moved across to the DIO to take on the appointment of Head of Army Technical Intelligence, which soon led to Head of Weapon Systems, Defence Intelligence.

.

30 WHERE TO FROM HERE?

In 1989, many things were coming to a head. Lo and I always said that we would put children before our careers, that we would never separate the family. The problems associated with frequently moving children that were growing up was looming to become a major decision factor in our lives. Up until now, carting the family all over the place, living in house after house, moving schools and leaving friends behind, had become a way of life that we had adapted to. After all, Lo and I had lived in 14 different houses over 24 years, more than half of those with at least one child in tow. It had made us a strong family unit. We were independent and our children capable within themselves. I'd been able to follow my military career, accepting postings and partaking in operations, exercises, conferences and adventures without limitation, knowing full well that I had a strong and capable wife, who ensured the family unit was solid... 'together'. Lo loved the Army life we led. She always looked forward to the next adventure, the next challenge, the next unknown. But our lives were progressing and the family unit was evolving.

Our children were already in, or approaching, their teenage years and it was obvious to Lo and I that we would need to make some decisions. The options were straight forward. They were not difficult to identify because many of my peers were going through the same dilemma and were faced with making similar decisions. What would be more difficult was the execution.

If I stayed in the Army we faced the choice between staying together, sending the kids away to school, or accepting 'service separation', where I would be separated from the family. By staying together I could soldier on, progress my career, accept new postings and we would continue to cart the children around with us. This would mean our children would have to face all the challenges that teenage children encounter as they grow up, with an added layer of difficulty. We'd be asking them to leave their friends and the stability of a known school environment behind every couple of years or sooner, for yet another move into yet another new environment. Alternatively, I could soldier on and Lo could come with me and we could put our children into boarding schools. This would separate the family and strain the special bond we shared, not to mention strain the budget. But, we always said we didn't want to separate the family. A third option was to leave Lo *and* the children behind as I was posted about the place. This would be recognised as 'unaccompanied postings' by the Army and there were conditions associated with such cases, where various travel allowances facilitated the posted person returning to the family frequently. Depending on the locations, this may be weekly on weekends, or monthly or quarterly. This option would also separate the family.

We had already watched other families go through the trials of 'service separation'. We watched families who'd sent their children to boarding schools, and we watched as some of the families carted themselves all about the place to stay together. We knew that relationships between husbands and wives are uniquely their own, and that each family is different, with each child responding to each situation in a different manner, but there is one common element... risk! The options available all presented risk to the stability of the marriage and the well-being of the family unit. None of the options seemed to be very palatable. We were not willing to take the risk.

That left the final two options: refuse to move or resign. I could 'hunker down' and refuse to be posted from the area or city of our choice, and where we believed our children would do best. I had seen others do that.

In my mind, that was not an option. Once an officer was identified as one who refused every posting unless it was in the city or area preferred, the career progression went only one way... down! The Army would begin to post that officer sideways into progressively less important and less challenging postings—soon everyone would know that officer had reached his or her 'rank limitation' and was not promotable: 'passed over' was the term used by others. This would lead to the inevitable: a disillusioned and embittered person. That was not for me. I had to love it or leave it.

So, that led to the final option. Resign!

The idea of leaving the Army had not seriously occurred to me before this last posting to DIO in Canberra. However, what else was there to do? I loved the Army: the work, the culture... the mateship. As a Lieutenant Colonel, I might be selected to command a regiment, and that would be fun. But that would mean a move to Western Sydney, Brisbane or Townsville and a heavy commitment in terms of time. Much more time away on duty, and a very busy social life. My predecessor, John Cox had been posted to a staff officers job in Army Headquarters. He was working for Major General Mueller, in an area called 'capability planning', a place where selected officers were posted. Mueller was a well-known senior officer, a very fine strategic thinker. My 'postings' people at the Military Secretary Branch told me that I was to be posted to work for Mueller when John moved on. General Mueller had personally approved my posting, and that alone was a significant 'feather' in my cap. John Cox was to become CO of 8/12 Medium Regiment in Sydney, and I was beginning to see a progression of career here for me also. So, if I chose to stay, my future in the Army looked good.

In the meantime, Lo and I had grown to like the Canberra area. We'd already decided that we would make a home around Canberra somewhere because Matt, Rae and Rob were doing well at school, all the facilities for sport and recreation looked fantastic and the city of Canberra appealed to us. It was not at all like we had been led to believe by many Army friends who had been posted there before us—a bit sterile and lifeless.

Beautiful tree-lined streets leading to flower-filled roundabouts, with generous parks surrounding a magnificent lake ringed by well-designed national institutions, placed there for our enjoyment. We loved it! And so, we bought some land at Braidwood, about 100km away. The plan was to build a house on that land and for the family to commute daily. Lo had by now secured a fulltime teacher's position at Lyneham High School in Canberra, so that would mean the five of us, in and out of Canberra from Braidwood every day. We bought a little wooden house, an ex-Navy married quarter of all things, that needed to be removed from its location on the Naval Communications site at Belconnen, and had it shifted to the land at Braidwood. Only just in time, because our lease at Hall, where we'd been living for the first two years, was just about up and we needed a place to live.

Our little house at Braidwood stood lonely and windblown on a hill on our land. It was barely a shell with so many gaps that blowflies and mice wandered in and out freely! It was winter... and it was cold. We used newspapers to plug the draughty holes in the walls! We piled into the car every morning and got home late each night, then started early again the next day. The pressure on the family was huge! Our little house on the hill needed a major upgrade and extension and I was planning to do it, but when... and how... and with what money? We had only just started to recover financially from our Australia trip.

I was tempted. If I left the Army, our children would be able to grow up in Canberra, in a stable school environment. I would get my pension, a part of which I could commute and get in cash, sufficient for building the house! My Defence pension would supplement Lo's income as a teacher, which would probably suffice. It was time. Being near Canberra would also enable me to pick up quite a bit of project work as a Reserve Officer. Reservists were paid tax-free hourly rates plus a travel allowance: a good plan, we reckoned.

I thought long and hard. This was a radical move. Lo was wonderfully supportive, ensuring I knew that she was happy either way, and that a decision to leave the Army was something I had to arrive at myself.

In the end, in 1991, I decided to resign from the Regular Army and transfer to the Reserve Army. Like many decisions in my life, it came quickly and easy, and I had no regrets.

I was excited at the prospect of being able to dedicate most of the time to house building and farm development. We had decided that we would become raspberry farmers. Of course, that's an obvious choice for an Army officer! What did either of us know about growing raspberries? Nothing! A sound basis for our farming decision! By chance we heard about a part-time raspberry farmer near Bungendore, about halfway to Canberra, wanting to sell up. We bought his goods and chattels and plants for one thousand dollars and confidently, with a lot of bravado and very little knowledge, set off to become farmers. Loaded with bags of raspberry canes, a few old freezers and a head full of advice, our lives had new purpose.

I made my resignation official. No qualms, no second thoughts. In went my letter to the Military Secretary! A few of my Army friends were incredulous.

'You're giving up your military career to become what? A farmer? You're an idiot, Lans!'

I can hear them, still. I did have a small twinge of doubt though, just before my final weeks, when another of my Lieutenant Colonel friends gave me a call. It was Tony Ralph, my classmate from Portsea and one of that gang of Queenslanders with whom I shared a train journey all the way from Brisbane to Melbourne, back when we were innocents. He was working for the Military Secretary at the time and knew everything there was to know of the careers of nearly all the Army's officers, including their peer rankings and future postings. Also, he knew where they sat in the promotion stakes.

He rang me:

'Ben, I just saw your letter of resignation. Come on over to my office mate, I have something to show you!' he said.

I had an inkling that he was going to try and dissuade me from my decision. He nearly did!

'Take a look at this list mate,' he said. 'This is the list of Lieutenant Colonels in the Regular Army today. There are about 120 of them. They're ranked in order of merit according to the score awarded each person as a result of their annual assessments. As you can see, you're well and truly in the top 20, in fact you're nearly in the top ten. This coming year, about 20 to 25 will be promoted to full Colonel. Now, you are not yet in the cohort that will be considered, but you're well set up for consideration in a few years' time if you don't stuff up. You sure you want to get out?' His voice trailed off with that all-important question.

I must admit that threw me for a time. I discussed it again with Lo. With this information, the choice and 'quality' of future postings was promising. Another overseas posting might be in the offing, such as Military Attaché in some exotic location for example, who knows? Equally there would also be the difficult postings that would come with the higher rank, the time consuming and more 'political' postings. Political in the sense of positioning yourself in the right place, time and sphere of influence in the race for a 'one star' rank: Brigadier and above.

But no, we were excited at the prospect of our new life. It was all go. Out into the big bad world I would go, where my old status and rank meant nothing, where there was no orderly room to manage my records, no free medical, no automatic pay increases (in fact no pay at all) and nobody to command but myself. There would be none of the 'boys' around to have an adventure with.

Or so I thought.

As it turned out, within days of resigning I was asked to come and do some work by another of my friends, Lieutenant Colonel Derek Leslie, who was the Director of Army Operations at the time. The Defence Department was in the process of acquiring new training areas, or as the press likes to refer to as 'ranges', which are large spaces of land, air and sea that stretch to hundreds of thousands of hectares. There was nobody available in the Army to project manage the implementation, the planned use and the design of these new training areas and the facilities that would be built on them. The Army had environmental teams, land acquisition experts, legal advisors, economic managers, strategic development officers, engineering managers, facility planners and builders... but no-one to tell them how that new piece of land, sea and airspace should be used. Where would the 'live firing' occur? Where should the manoeuvre corridors best go? Where should we put the airstrips, the camps, the roads? The questions went on. What does a new training area look like?

'How should I know?' I said to Derek with a shrug.

'You're a gunner mate! You have used training areas all your life! You would know better than most!' He answered with a grin I can still remember today.

Well there it was. I was only just out of uniform as a regular soldier only to step back into the same uniform as a reserve soldier. I planned to do two to three days per week on my reserve job, leaving plenty of time for the house and farm. This could be fun I thought, and I can be quite useful in this role.

It *was* fun. I *was* quite useful. The reserve job lasted a few years and led to my becoming a Defence consultant from 1999 onwards, providing well paid professional services. These included the role of training area development specialist, strategic planner, project manager, project management system trainer and simulation training developer. I managed and wrote studies, reviews, development papers and Defence White Paper contributions that directly led to over a billion dollars of Defence investment. I was fortunate to travel the width and breadth of America including Alaska and Hawaii, to learn about modern training methods and how they could benefit Australia, and to host the Americans here... taking them all over Australia and its outback, to plan future joint, combined training.

As a consultant it was an honour to be asked to provide input to Defence White Papers on the development of training facilities in Australia; it was a surprise to be asked to facilitate the planning of the Air Force's principal Air Weapons Ranges; and it was a pleasure to be conducting briefings about training area facilities, at Vice Chief of the Defence Force and Chiefs of Service level. More significantly, and this was the key to achieving successful outcomes, I was humbled to be accepted as an equal by serving military officers of all ranks: smart young 'fast jet' pilots who had flown dangerous hostile missions in Iraq; and calm, intense Army commanders who had just returned from the dirty war that is Afghanistan. They saw me as a partner, one who identified with, and quickly understood, their skills and requirements and the military science associated with their unique, specialised roles in war, and how to apply that to training in Australia.

Who would have thought that I would have the opportunity to back up a satisfying military career with an equally satisfying consulting one, all within the same Defence force? The part-time hours turned into a full-time career from 1999, when I left the reserve Army, until 2017. It significantly helped Lo and I enrich our family life and manage our financial stability. It brought adventure and travel and dealings at the highest levels of Defence.

But that is another story.

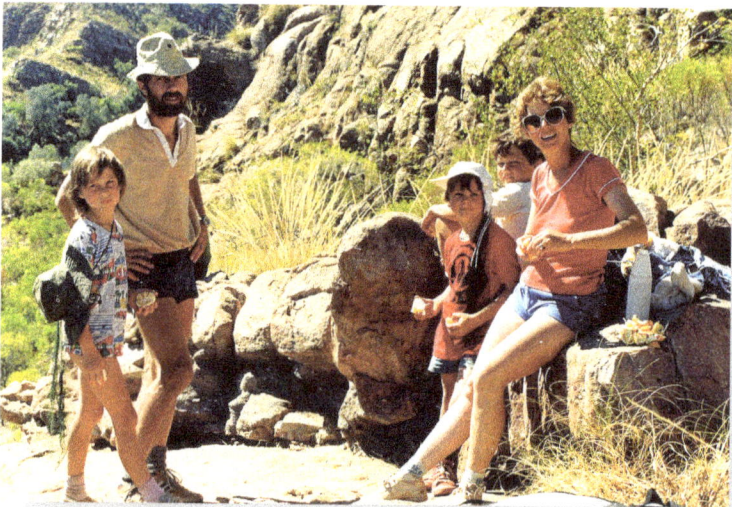

The family on our Australian adventure after coming home from the London posting. With Lo and I are Rae on the left, then Rob in the middle and Matt at the back.
Photo Author

ACKNOWLEDGEMENTS

There is one special person without whom most of this memoir would not have happened: Lo, my wife, the love of my life and my staunchest supporter. She was there through good and bad, always the strong, reliable team member, ever encouraging me in my career and ready to be posted wherever I was sent in the world, where she would instantly set about making a home for our family. When I wanted to seek even more adventure with my men, and explore wild places, she let me go with love in her heart, never for a moment admitting to me how lonely and hard it was to be home alone with the kids so often.

Then, when I wanted to tell the story of my memoirs, she was right there, encouraging me, laughing at the funny bits, bringing me to some reality when I got lost in the story telling, cajoling me into continuing and helping with edit after edit to make it an enjoyable, and hopefully meaningful read, especially for our children. Lo, I can barely express how lucky I am that you shared not only my journey, but also the telling of it.

A special thanks to Lo's parents, who acted as sounding boards and were subjected to many reads and re-reads and whose comments and enjoyment were infectious. Sadly my own parents died early in my life and were unable to enjoy this story. It was my father who inspired me to join the Army and my mother whose love of nature inspired me to choose an outdoors type of career.

Unfortunately my brother Reg, who was born twenty years before me, passed away before this book was published. However, he was able to read most of the manuscript, his wife Anna reading him the final chapters days before he died.

A big thank you to Trina and Sally, who gave their time to edit my work, and Leslie, who encouraged me to publish and gave me lots of good advice.

There are many Service colleagues and friends, in the Australian Forces, as well as American and British Forces, who shared my experiences and adventures and who gave me counsel along the way. I have tried to tell our tales with humour and honesty. I thank them all for their professionalism and their mateship but in particular I thank the men of 108 Field Battery, who served in the period between 1977 and 1981, with whom I share a special bond, never to be broken.

Finally I would like to acknowledge John Guy, whose voice was so inspiring early that morning, my first day in the Army. It gave me confidence and comfort just when I needed it.

www.ingramcontent.com/pod-product-compliance
Lightning Source LLC
Chambersburg PA
CBHW041143230326
41599CB00039BA/7147